Adding Space

Without

Adding On

 CREATIVE HOMEOWNER PRESS

Writers: Jane Cornell, Mark Feirer, David Jacobs
Editorial Director: David Schiff
Senior Editor: Timothy O. Bakke
Associate Editor: Patrick Quinn
Copy Editor: Beth Kalet
Photo Researcher: Jane Cornell

Art Director: Annie Jeon
Graphic Designer: Michelle D. Halko
Illustrators: Craig Franklin, Ed Lipinski, James Randolph,
    Paul M. Schumm, Ray Skibinski

Cover Design: Annie Jeon
Cover Photograph: Bill Rothschild/Ginsburg Development
    Corp. "Boulder Ridge"

Electronic Prepress: TBC Color Imaging, Inc.
Printed at: Webcrafter Inc.

Current Printing (last digit)
10 9 8 7 6 5 4 3 2 1

Adding Space Without Adding On
Library of Congress Catalog Card Number: 96-84696
ISBN: 1-880029-85-5

Photo Credits

p. 1: Eric Roth Photography, Boston, MA/Faye Etter

p. 6: Phillip H. Ennis Photography, Freeport, NY/Paul Erdman, Architect, Oyster Bay, NY

p. 9: Elizabeth Whiting Associates, London, UK

p. 10 (top l): Bill Rothschild, Wesley Hills, NY/Ginsburg Development Corp. "Boulder Ridge"

p. 10 (top r): Melabee M Miller, Hillside, NJ/Andrea Longo, Interiors, Toms River, NJ

p. 13 (top): Melabee M Miller/Lois Caruso, ASID, Lois Caruso Interiors, Red Bank, NJ

p. 13 (bot.): Nancy Hill, Mt. Kisco, NY/Roy Barnhart, Fairfield, CT

p. 14 (top): Nancy Hill/Brassard Design Associates, Red Bank, NJ

p. 14 (bot.): Bill Rothschild/Bob Goldstone

p. 16: Melabee M Miller/Norma Hayman, Interiors, Rumson, NJ

p. 17 (top l): Bill Rothschild/Ginsburg Development Corp. "Winchester Village"

p. 17 (top r): Bill Rothschild/Denise Balassi

p. 17 (bot.): Bill Rothschild/Denise Balassi

p. 18: Bill Rothschild/Beverly Ellsley

p. 19 (top): National Kitchen & Bath Association (NKBA) Design Competition/Gay Fly & Cris Baker, Gay Fly Designer Kitchens & Baths, Houston, TX

p. 19 (bot.): Melabee M Miller/Mikel Patti, ISID, Surroundings, Manahawkin, NJ

p. 20: Ted Harden, New York, NY

p. 22 (top): Ted Harden

p. 22 (bot.): NKBA Design Competition/Gary S. Johnson, International Kitchen & Bath Exchange, Sunnyvale, CA

p. 23 (top): Maura McEvoy/NKBA Design Competition/Kaye Kingsland & Scott Lipp, Innovative Cabinetry, Boca Raton, FL

p. 23 (bot.): Melabee M Miller/Victor Moldovan, ASID, VLM Design, Island Heights, NJ

p. 24 (top): Lindal Cedar Homes, Seattle, WA

p. 24 (bot.): NKBA Design Competition/Frank Diliberto, Huntington Kitchen & Bath, Huntington Station, NY

p. 25 (l): Wood-Mode Fine Cabinetry, Kreamer, PA

p. 25 (r): Bill Rothschild/Ed Singer

p. 26: Maura McEvoy/NKBA Design Competition/William Earnshaw, A&B Kitchens & Baths, Wyckoff, NJ

p. 28 (top): Bill Rothschild/Frank Lavin

p. 28 (bot. l): NKBA Design Competition/Willi___ & Baths, Wyckoff, NJ

p. 28 (bot. r): Bill Rothschild/Ginsburg De___

p. 30: Nancy Hill

p. 32 (top): Velux-America, Greenwood, ___

p. 32 (bot.): Bill Rothschild/Lee Napolitan___

p. 33: Elizabeth Whiting Associates

p. 45: Elizabeth Whiting Associates

p. 61: Elizabeth Whiting Associates

p. 79: Nancy Hill/Robert Davis, AIA, Edina, MN

p. 99: Phillip Ennis Photography/Beverly Ellsley Interiors, Westport, CT

p. 113: The Terry Wild Studio, Williamsport, PA

p. 139: Elizabeth Whiting Associates

p. 159: Phillip Ennis Photography/Ronald Bricke & Associates, New York, NY

p. 175: Bruce Hardwood Floors, Dallas, TX

Saftey First

Though all the designs and methods in this book have been tested for safety, it is not possible to overstate the importance of using the safest construction methods possible. What follows are reminders; some do's and don'ts of basic carpentry. They are not substitutes for your own common sense.

▶ *Always* use caution, care, and good judgment when following the procedures described in this book.

▶ *Always* be sure that the electrical setup is safe; be sure that no circuit is overloaded, and that all power tools and electrical outlets are properly grounded. Do not use power tools in wet locations.

▶ *Always* read container labels on paints, solvents, and other products; provide ventilation, and observe all other warnings.

▶ *Always* read the tool manufacturer's instructions for using a tool, especially the warnings.

▶ *Always* use holders or pushers to work pieces shorter than 3 inches on a table saw or jointer. Avoid working short pieces if you can.

▶ *Always* remove the key from any drill chuck (portable or press) before starting the drill.

▶ *Always* pay deliberate attention to how a tool works so that you can avoid being injured.

▶ *Always* know the limitations of your tools. Do not try to force them to do what they were not designed to do.

▶ *Always* make sure that any adjustment is locked before proceeding. For example, always check the rip fence on a table saw or the bevel adjustment on a portable saw before starting to work.

▶ *Always* clamp small pieces firmly to a bench or another work surface when sawing or drilling.

▶ *Always* wear the appropriate rubber or work gloves when handling chemicals, doing construction, or sanding.

▶ *Always* wear a disposable mask when working around odors, dust, or mist. Use a special respirator when working with toxic substances.

▶ *Always* wear eye protection, especially when using power tools or striking metal on metal or concrete; a chip can fly off, for example, when chiseling concrete.

▶ *Always* be aware that there is seldom enough time for your body's reflexes to save you from injury from a power tool in a dangerous situation; everything happens too fast. Be alert!

▶ *Always* keep your hands away from the business ends of blades, cutters, and bits.

▶ *Always* hold a portable circular saw with both hands so that you will know where your hands are.

▶ *Always* use a drill with an auxiliary handle to control the torque when large-size bits are used.

▶ *Always* check your local building codes when planning new construction. The codes are intended to protect public safety and should be observed to the letter.

▶ *Never* work with power tools when you are tired or under the influence of alcohol or drugs.

▶ *Never* cut very small pieces of wood or pipe. Whenever possible, cut small pieces off larger pieces.

▶ *Never* change a blade or a bit unless the power cord is unplugged. Do not depend on the switch being off; you might accidentally hit it.

▶ *Never* work in insufficient lighting.

▶ *Never* work while wearing loose clothing, hanging hair, open cuffs, or jewelry.

▶ *Never* work with dull tools. Have them sharpened, or learn how to sharpen them yourself.

▶ *Never* use a power tool on a workpiece that is not firmly supported or clamped.

▶ *Never* saw a workpiece that spans a large distance between horses without close support on either side of the kerf; the piece can bend, closing the kerf and jamming the blade, causing saw kickback.

▶ *Never* support a workpiece with your leg or other part of your body when sawing.

▶ *Never* carry sharp or pointed tools, such as utility knives, awls, or chisels, in your pocket. If you want to carry tools, use a special-purpose tool belt with leather pockets and holders.

Contents

Introduction 6

1 Planning and Design 9
Costs: Money, Mess, Time, and More 10
Laying Out Floor Plans 11
Utilizing Unused Areas 14
Planning Rooms 20
Lighting 30

2 Sizing Up the Project 33
Surveying the Attic .34
Surveying the Basement 36
Surveying the Garage 40
Planning for Utilities 40
Who Will Do the Work? 42
Abiding by Building Codes 43

3 Preparation Work 45
Planning the Logistics 46
Wall Removal 47
Removing a Non-Bearing Wall 49
Garage Door Removal 51
Joists and Rafters 53
Installing Attic Ceiling Joists 54
Concealing Heating Systems 55
Moisture Problems .55
Sealing a Masonry Wall 56
Installing a Sump Pump 59

4 Building Stairs and Framing Floors 61
Basement Stairs 62
Attic Stairs 63
Calculating the Rise and Run 63
Building the Stairs 64
Attic Floors 67
Measuring a Floor for Loading 68
Installing Subflooring 71
Basement Floors 72
Repairing Cracks in a Concrete Floor 72
Installing an Insulated Subfloor 73
Garage Floors 75
Closing the Foundation Wall 75
Building an Elevated Subfloor 77

5 Attic Framing and Dormers .79
Building Partition Walls 80
Using the Tip-Up Method 80
Building a Wall in Place 82
Building a Sloped Wall 82
Kneewalls 84
Insulation and Ventilation 86
Insulating an Attic Roof 87
Adding Rigid Foam .88
Dormers 89
Planning the Dormer 91
Building a Shed Dormer 92
Building a Gable Dormer 96

6 Basement and Garage Framing 99
Fastening Objects to Masonry 100
Building Partition Walls 102
Building a Tip-Up Wall 102
Building a Wall in Place 104
Insulating Masonry Walls 105
Insulating with Fiberglass 105
Installing Rigid Insulation 108
Enclosing the Garage Door Opening 110

Too often, home-conversion efforts end in frustration and failure because the owners moved too fast, without realizing that it was going to take longer than expected to finish certain jobs or that they were going to need more material than expected and went over budget. Give yourself plenty of time to plan the overall project and visualize how you'll accomplish the tasks that lay ahead.

Before you start the project, consult with your local building department to determine which permits are needed and how you're supposed to request the mandatory inspections by the building-department officials. Most do-it-yourselfers who have dealt with building-department officials in a reasonable fashion have walked away with more building knowledge than they had before they walked in the door. Don't be afraid to ask questions. And of course, keep this book handy for reference.

Tools and Materials

This book takes you through the logistics of converting unused spaces in your house into useful, beautiful living areas—from planning the projects to constructing and finishing the space. The first two chapters help you make basic decisions about what to do with the untapped potential in your house; the last eight tell you exactly how to implement those decisions. Each how-to section begins by rat-

ing the level of difficulty of the task at hand. The level of difficulty is indicated by one, two, or three hammers:

Easy, even for beginners.

Moderately difficult, but can be done by beginners who have the patience and willingness to learn.

Difficult. Can be done by a do-it-yourselfer, but requires a serious investment in time, patience, and specialty tools. Consider hiring a specialist.

A list of tools and materials required for the work is also provided with each how-to section. Certain hand

tools are basic to most of the construction and remodeling tasks described throughout the book:

BASIC CARPENTRY TOOLS

▶ Combination square
▶ Flat-bladed screwdriver
▶ Hammer
▶ Handsaw
▶ Measuring tape
▶ Pencil
▶ Phillips screwdriver
▶ Pliers
▶ Plumb bob
▶ Pry bar
▶ Spirit level
▶ Utility knife

To work on plumbing systems, you'll need other basic tools:

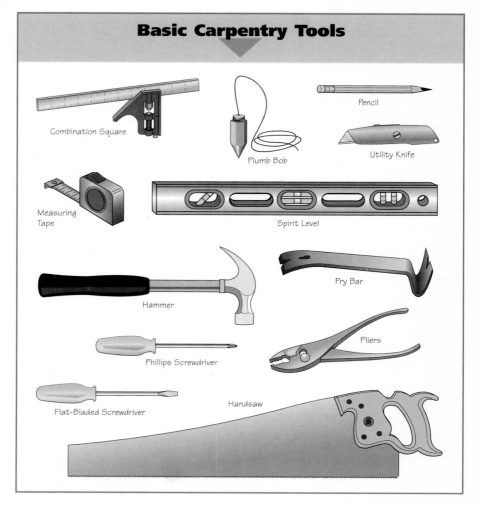

Basic Carpentry Tools

Combination Square

Plumb Bob

Pencil

Utility Knife

Measuring Tape

Spirit Level

Pry Bar

Hammer

Pliers

Phillips Screwdriver

Flat-Bladed Screwdriver

Handsaw

Basic Plumbing Tools

Basic Electrical Tools

BASIC PLUMBING TOOLS

▶ Adjustable pliers
▶ Emery cloth
▶ Hacksaw
▶ Leather gloves
▶ Measuring tape
▶ Propane torch
▶ Round file
▶ Tubing bender
▶ Tubing cutter

Almost all electrical work can be accomplished with the following assortment of tools:

BASIC ELECTRICAL TOOLS

▶ Cable staples
▶ Electric drill and assorted bits
▶ Electrician's tape
▶ End nippers
▶ Hacksaw
▶ Hammer
▶ Measuring tape
▶ Needle-nose pliers
▶ Plastic wire connectors
▶ Screwdrivers (flat-bladed and Phillips)

▶ Utility knife
▶ Voltage tester
▶ Wire stripper

Immediately following the tools and materials list for each project throughout the book, step-by-step instructions show how to execute the work. Follow the directions to complete the project successfully.

Time Factor

Lastly, do yourself the biggest favor of all—allow plenty of time to complete the project. If you expect that your attic, basement, or garage conversion will take two months to complete, give yourself three months to get the job done. Every professional home remodeler will tell you that each job inevitably winds up posing a problem or two that had never been considered. You may have to confront out-of-plumb walls, newly found ground-

water seepage problems, deteriorated structural members, and inferior workmanship done when the house was originally built. These obstacles and others like them are common, so if you come across one or two in your project, don't despair. Take a deep breath, roll up your sleeves, and fix the problem. Look at the situation as an opportunity to learn something new about home repair and remodeling, then pat yourself on the back for allotting those extra few weeks in your schedule to finish the project completely and professionally.

While the challenges are great, the rewards are even greater in making an existing home work better for you. You can create a custom house without incurring the costs of new construction, making your home fully realize its potential and often adding to its market value. ▲

Planning and Design 1

A sometime cook becomes a gourmet guru. A newborn becomes a schoolboy and needs more space for studying, collecting, and pursuing hobbies. The single bathroom is totally inadequate. What was charming, cozy, and intimate before now feels like a cramped rabbit warren of restrictive rooms. If not one of these, you probably have other complaints about your home's space that are screaming for solution. However, don't focus solely on what's most obvious. Now is the time to consider all the needs to be met by your remodel, both currently and for the future. Only then can you really improve without adding on. Here's how to start.

▲ *Expanding into the attic space allowed room for a full luxurious master bath.*

▲ *In this attic remodel, low bookcases make efficient use of the knee-walls and create a cozy sitting room.*

Costs: Money, Mess, Time, and More

Cost is a compelling reason for forgoing a major addition. In most cases it's far cheaper to work with what you have, maintaining the original house footprint, than to make major additions. If you're considering an addition, you'll have to figure in the costs of excavating and pouring a foundation, erecting the framework, and closing in the structure—all before you begin any interior work. Obtain estimates for these jobs from at least one but preferably three contractors to get an idea of what the job will cost. Come up with a plan for the addition and one for a remodel of the attic, basement, and/or garage. Make up rough materials lists and bring them to the lumberyard or home center store. Be realistic and figure in an extra 10 percent for wastage and any mistakes in calculations. After you get the estimate for materials cost, you'll be able to decide upon reasonable and logical budgets.

If you still want to consider an addition, make a side-by-side comparison of the remodeling and addition budgets. Start the list with overall costs, then factor in the following considerations as well:

Overbuilding

If resale of your home is a possibility, or merely to keep peace with neighbors, make sure that you're not overbuilding with any planned additions. Ask local real estate agents for their impressions of the stability of your neighborhood and their recommendations for what the top dollar

would be for a house renovated as you envision. In most cases, you will recoup far less money if you build in value well beyond that of the surrounding homes. It's often financially more prudent to make your existing house more efficient, utilizing all possible spaces, than simply to add on. Even if additions are in your future plans, make the existing spaces work efficiently now.

Building Restrictions

Almost all neighborhoods have restrictions controlling where and what you can build on your lot. The

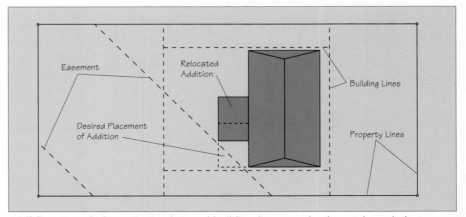

Building Restrictions. *Easements and building lines may lead to awkward placement of an addition, which is a good argument for remodeling.*

restrictions may include the amount of inhabitable space you can have, especially if you're in a planned community. Adding a new bedroom, for example, may increase the occupancy rate beyond what's acceptable. Other external features, such as building height or setback on all four sides, may be determined by the local building code.

There are the "good neighbor" considerations as well. Perhaps your addition would block the sunlight to a neighbor's yard, or your expansion may overwhelm the curbside appeal of another home. If you plan to live in the house for some time, make sure your additions won't make enemies of your neighbors.

Utility easements and the like that extend across your property may also drastically limit the amount you can expand your building. Reorganizing your home within its present shape, however, won't cause conflict with any easements.

Coping with the Mess

It's said that marriages are truly taxed during remodeling. Mess, disarray, clutter, workers at odd hours, unexpected delays, and even the need to move out for a time all take their toll. You need a realistic assessment of the amount of mess and time your remodeling will entail. Equally important is your family's tolerance level. Consider these pointers for both additions and remodels:

▶ Will you uncover dangerous substances such as lead paint or asbestos? How costly will it be to correct these problems?

▶ For below-ground areas, are they radon-free?

▶ Can you organize your remodeling in easy-to-live-with stages?

▶ If you use outside professionals, are they likely to stay on schedule?

▶ Can you isolate the remodeling from the remainder of the home, and at which stages of the project?

▶ Can you keep your home operational during the remodeling? Can you continue to use the kitchen and laundry areas, and reserve at least one bedroom as a refuge from the remodeling? If not, can you set up temporary camp-style living that's adequate?

▶ Which season is best for this operation? Should you plan your project for spring and summer, for example, when you can make use of your outdoor space?

▶ Will the remodeling disrupt your children's schooling if undertaken during the school year? Can you send the children away to visit relatives or to summer camp if the home is so disruptive that it would be best to have them gone?

▶ What steps can you take to ensure that children won't be exposed to danger during the remodeling? Can you close off an area to children and easily block the remodeling site?

▶ Is your current electrical service able to handle the added living space? Your heating system?

Laying Out Floor Plans

Know exactly what you have to deal with by making layouts of the rooms as they are now. You can purchase kits with grid paper and ready-made symbols of architectural details and furniture templates, or make your own using graph paper with a ¼-inch grid to create a ¼-inch- or ½-inch-per-foot scale for your layout.

Make sure you have a 25- or 30-foot 1-inch-wide tape measure and a scale ruler, which is marked for drawing plans to scale. Use the tape measure to determine the exact dimensions of closets, counters, furniture, desks, appliances, and other things you expect to put into your new living space. Draw the items to scale on your plans to see how much maneuvering room you'll have after the new conversion is furnished. Take your time and openly discuss plans with other family members; they may have some great ideas for the conversion that you never thought about.

Give yourself a bird's-eye view on paper of your house plan. The plan view will help you anticipate how changes in one room will affect other areas. At this point, determine which are the load-bearing walls—usually exterior walls and some interior walls. Bearing walls support floor joists or rafters. Generally, bearing walls run in the same direction as the subflooring: Joists rest on the load-bearing walls crosswise, and subflooring panels or boards rest on the joists crosswise, so they run parallel with the walls. While bearing walls can be eliminated if you provide adequate alternative support, you may want to avoid the cost of changing them. Get expert advice from a building contractor or engineer before tackling this redesign unless you absolutely know what you're doing.

Next, cut out templates to represent your furniture. Mark the pieces so you know what's represented—a chair, desk, buffet, and the like. Back the icons with stick-and-release glue so you can reposition them easily and securely. Position the furniture as you now have it. Trace or photocopy your layouts so that you can redesign and make changes without marking up the originals.

Add Traffic Patterns. Next, draw in the traffic patterns you ordinarily use in moving throughout the house.

EXISTING FLOOR PLAN

FIRST REDESIGN

SECOND REDESIGN

Laying Out Floor Plans. *What seems an ideal arrangement may improve once you give more thought to your plan. Reassigning room use and rearranging space in the initial plan leads to a better floor plan with fewer structural changes.*

FLOOR PLAN

EXISTING TRAFFIC PATTERN

REDESIGN

Add Traffic Patterns. *By laying traffic patterns into the floor plan of a house, you can see which areas are under-used and where redesigning is required. In the plan showing existing traffic patterns, the living room is obviously under-used, and the family required a fourth bedroom. The redesigned floor plan better complements the family's living style.*

While you already may know the major traffic problems, you might discover others that aren't as bad but are nonetheless annoying. Perhaps you keep outdoor clothes in an entry closet near the front door but usually use the back door at the other end of the house. You'll see that a mudroom or closet near the back door would be a timely addition. Ask all family members to add to the traffic patterns, even doing their own traffic patterns in their own colors. Common problems include

▶ Traffic paths through the kitchen's cooking area.

▶ Traffic paths through conversational furniture groupings instead of around them.

▶ Doors that bump into one another.

▶ No adequate storage for activities that demand them.

▶ Rooms that must be reached through another room such as a bedroom.

▶ Disproportionate use of space, with some rooms rarely used and others constantly crowded.

Establish a Wish List. Traffic may not always be the problem. Incorporate other needs you want to fulfill as well. High on the "most wanted" list for many homeowners is more storage, easier access to a space, increased openness and natural light, and areas that provide privacy. Obvious answers are to incorporate a basement, garage, or attic for living space. But the best solutions also call for a rethinking of the other rooms.

▲ Unless you'd been here before the remodel, you'd never know this formal living room was a garage conversion.

▲ A daylight walk-in basement, with its easy access to natural light and the outside yard, makes for a delightful informal family sitting and gathering room.

▲ *Enclosing a porch creates additional living space and maintains a casual atmosphere. In this instance, painting the original exterior brickwork adds interest.*

▲ *A garage became a second living room while the breezeway to it was expanded and enclosed as a sunspace that unifies the house.*

Making Use of Unused Areas

If you have an unfinished attic or basement or an attached garage, your search for more space is over. These are the most obvious and popular areas for finishing. Note that in converting an attic, basement, or garage to living quarters, you're transforming space that wasn't designed for the kind of general access you'll require once the project is done. Some modifications may be required to make the new living space easily accessible. Other candidates for conversion include porches and breezeways. And if you're fortunate enough to have a two-story space, as in a garage with an attic space, you can co-opt some of that overhead space as a loft.

Gaining Attic Access

Few houses are designed with an attic conversion in mind, so access routes are typically rudimentary. In some cases, the only access is through a hatch plate tucked into the ceiling of a closet. Other houses may have pull-down stairs, but these can't legally be used to reach a finished attic. Because a pull-down stair fits into an opening that's only about 26 inches wide, the opening has to be enlarged to accommodate the width of a standard stairway. A standard, straight-run stairway is about 36 inches wide and 11 to 13 feet long without landings. You must allow at least 36 inches for a landing at the top and bottom. Building codes require a minimum vertical clearance of 80 inches at all points on the stairs. If you need to add a stairway to the attic, subtract the floor space it will take from your usable total, both in the attic and on the floor below.

Locate the Stairway. First make two decisions: Where will the stairway start and where will it end? Look for under-used space below the attic. If a wall is removed from between two small rooms, the resulting larger room possibly could provide the needed space. A bedroom closet might be changed into a stairwell. You might even decide to sacrifice a small room to gain suitable access to the attic if doing so results in a net gain of floor space. Once you've found a starting place, you can determine where the stairway will end. Ample headroom at the top of the stair isn't always easy to find in a room that has ceilings sloping to the floor. Terminating the top of the stairs near the center of the attic provides the greatest headroom above. You can place the stairway closer to attic walls, however,

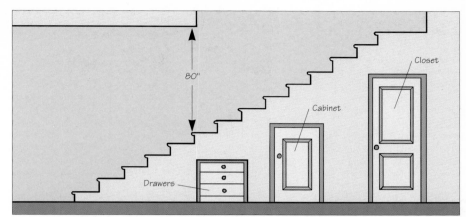

Gaining Attic Access. *Code requires that stairs have at least 80 in. of headroom measured vertically at the front edge of the steps. Where headroom is less than 80 in., the attic floor must be cut away. The stairwell can be used for storage.*

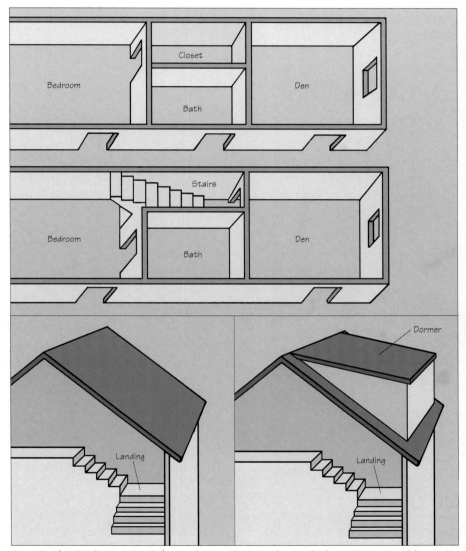

Locate the Stairway. *Look for under-used spaces beneath the attic, or consider changing room configurations. Here, a stair replaces a closet (top). To conserve space, arrange the stairs so that they descend as they step toward the eaves (left). You can move the stairs even closer to the eaves by building a dormer to gain headroom (right).*

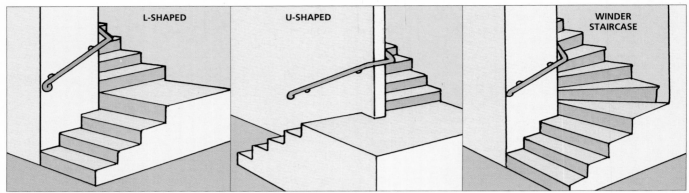

L- and U-Shaped Stairs. *These stairs are harder to build than straight stairs but don't require a long stretch of space. L-shaped (left), U-shaped (middle), and winders (right) must be code-approved.*

Spiral Stairs. *This is the most compact design but often is difficult for children and elderly persons to navigate. Spiral stairs may not be allowed in your area, so check local building codes.*

▲ *A charming playroom in this attic has a painted floor and a picket fence with stenciled flowers that echo wall angles.*

if its angle follows the angle of the roof. Another trick is to build a dormer over the stairwell. A dormer not only provides suitable design, it also provides the stairs with plenty of natural light.

L- and U-Shaped Stairs. If there's not room for a straight-run stair, consider L- or U-shaped stairs. Though they're harder to build, these stairs are more compact and don't require

the length of uninterrupted floor space needed by straight-run stairs.

Spiral Stairs. From a purely visual standpoint, there's nothing quite like a set of spiral stairs. They can be installed in a space as small as 48 inches in diameter but may be difficult for some people to use (particularly the elderly). Also note that it's challenging, if not impossible, to get furniture up and down

spiral stairs. An advantage of using spiral stairs, however, is that they can be purchased as a kit and assembled on site rather than built from scratch. Check your local building codes; some restrict the use of spiral stairs.

Uses for Attic Conversions. Attics are ideal for establishing a master suite. The space at the top of the house lends itself to the serenity of a bedroom, and it's usually not

▲ *A triple window dormer creates additional space for this children's room while providing cheerful light.*

▲ *A child's room takes advantage of built-ins in this attic conversion that provides room to sleep, study, and play.*

difficult to tie into the plumbing system for the master bath. A large attic can also be divided into two or more children's bedrooms with an accompanying bath. A home office also makes good use of the solitude and quiet of the attic, as long as you don't require a separate entrance for visiting clients and the like. Other common uses for attics include family or sitting rooms and entertainment centers, as long as they don't generate a lot of loud noise and activity. Areas designed for active play, especially among older children and preteens, are out of the question.

Gaining Basement Access

Access to basements isn't easily altered without a great deal of remodeling work in other parts of the house. Basements are generally entered through stairways located near kitchen spaces. Unless there's an easy way to position a new stairway through an unused space like a closet or empty bedroom scheduled for remodel, you may have to keep what you have. Should you have a special reason for wanting to change the location of an interior basement

stairway, consult with a building engineer or architect. The means by which a new access opening is made will most likely require serious structural changes in the original house design. The same goes for an existing exterior access. Any new point of entry will require cutting out a section of a foundation wall. This could prove costly, as the new opening will have to be structurally reinforced and the old opening securely sealed. It would be more advisable to alter the interior basement floor plan to make better use of the existing access point. If yours is a daylight or walk-in basement, where one wall is completely exposed, you can create the look of an above-grade room.

Uses for Basement Conversions.
Virtually all the functions recommended for attics can be translated for basements. Additional bedrooms, a children's room and/or playroom, a family room, an exercise

area, and a home office are possibilities for most basements. Basements are especially suited to uses that are related to activities. Home crafters and hobbyists often enjoy having a basement work space where mess isn't transferred into the rest of the house. Those in the needle crafts can set up workshops and sewing rooms adjacent to laundry areas, capitalizing on the synergies between these two activities. Home handymen often prefer to set up shop in the basement. And for true abandon, kids love having a play space that's set up with easy-care, no-fuss materials for roughhousing.

▲ *A media wall is in full view of both casual seating and exercising equipment in this basement remodel.*

In daylight basements, you often can create a second self-contained "apartment" for a relative or for use as an extra income generator (if zoning allows). Separate entrances, since the basement is on ground level, complete the sense of privacy. In below-ground basements, should you want to add a kitchen or bath, you can tie into the plumbing system that's in place above. You can buy toilets and lavatories that are specifically designed for below-grade applications.

Gaining Garage Access

The location of an exterior garage access opening at grade level is easy to seal off and relocate. Depending upon your house's interior floor plan, access from the garage conversion to the house could be changed to serve your needs best. Since the garage wall is most likely framed with wood, you can cut in a new opening to serve a proposed office space or master bedroom better. You may want access to a home office in the garage conversion from a front door foyer, rather than from a kitchen. You must be aware of which walls are load-bearing, though, because you'll have to erect a temporary support while remodeling the wall.

Uses for Garage Conversions. Garages are untapped sources of space that are usually easy to convert. What matters most in determining a garage's best function is how it is integrated with the rest of the house. If the garage is just off the kitchen with a doorway already in place, for instance, it's an obvious choice for a family-room expansion. This location would not be so great for a master bedroom suite. Garages in many homes are at least two steps down from the adjoining areas. You'll want to make a smooth transition from one area to the garage conversion, especially if your intent is to open up the garage as a family room extension of the kitchen. For comfort in most northern climates, put in new floor decking rather than laying flooring directly on a concrete slab garage floor. This also raises the floor level somewhat. Make the lowered level of a garage conversion a decorative asset. Join the levels with dramatic

▲ An entire kitchen was moved into this converted garage, with original rafters maintained.

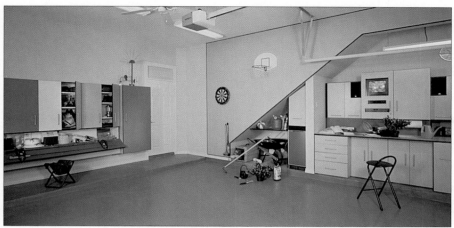

▲ *This garage retains its conventional use but has been designed to double as a work-room and playroom when cars are left outside.*

stairs. For safety make sure stairs are clearly demarcated by using different flooring on the varying levels and providing handrails.

Garages that are not attached can also be converted. An enclosed walkway can bind them to the main house. The separation itself may be perfect for a studio or home office, even a child's playroom. Since a garage is an open space that usually can be readily finished, ask yourself the same questions you would about finishing off a basement or attic. However, converting a garage has some unique advantages and special considerations.

Most garages function as additional storage areas for garden and sports equipment, off-season parapher-nalia, workshops, pantry goods, and whatever the family needs. Determine first how much room you'll still need for storage and whether there's enough room to convert once those storage needs are met. Check your local zoning regulations to see whether you can put up a shed for storage and devote the entire garage to conversion.

New Parking Areas. Be realistic about whether you want to park out in the open. It's not good for a car to be constantly exposed to snow or unrelenting sun. It's also nice to have a sheltered area when you want to keep the car in bad weather. Is there a space to accommodate a detached garage on your property? Consider also attaching a garage on the other side of the current garage or another part of the house. Perhaps a simple carport would suffice.

The Driveway. A garage conver-sion can appear awkward from the front of a house if the garage door is replaced with a wall while the drive-way is left intact simply to run into a wall. Rather than tear out the driveway, consider leaving the garage door in place and starting the conversion 6 or 8 feet back behind the garage door. Following this plan would give you a handy space for a workbench and storage of land-scaping tools, while also permitting the front of your house to look orig-inal as viewed from the street. Your garage con-version would be hidden from an outdoor perspec-tive. You can also try to use the driveway in relo-cating another garage or carport or just as a park-ing space. You might instead convert the drive-way into a patio adjoining the newly converted garage space or use it as the base for an outdoor patio or deck.

Garage Doors. At first glance, the obvious conversion for a garage door is to install patio or sliding glass doors, providing a transition of the indoors and out. Unfortunately, garage doors often overlook the street, providing no privacy. If you plan to park in front of the converted garage, that's no lovely vista, either. Solutions include finishing off the exterior of the garage, including the garage-door opening, to match the rest of the house. Then you can place new windows and doors where they'll overlook an attractive view like the garden. Screening off a small area outside the garage may be the most economical and satisfying solution. Use trellis fences and the like to form a background for a miniature garden and protect the converted garage from street view.

Lofty Ideas

A library, small office, study area, secondary guest room, play area

▲ *A stair landing has been claimed for a library corner in this remodel.*

▲ *A loft over a two-story living room becomes an entire room unto itself.*

adjoining children's rooms, or den are just some of the functions compatible with a loft. You may be able to reach a new loft through an adjacent room on the upper level. This takes up the least amount of floor space. If you must add a stairway, plan its location carefully to rob the least from floor area both below and in the loft itself. Some contemporary lofts are reached across walkways that span open spaces below. Other options include extending a walkway out from the loft itself to locate a stairway where it would be most convenient below. A circular stairway may be an ideal solution, and take the least floor space of all.

Safety. Barriers to the loft edge are critical in loft design, especially with small children around. Use balustrades and banisters to provide an open look. A wainscot half-wall 30 inches or more high provides a solid

barrier. Make sure you don't need local approvals to add a loft. As in building a deck, you need to make sure that the loft footings or other support are adequate to the load and solidly supported.

Aesthetics. You can easily work a loft area into many architectural design styles. For instance, you can use rustic rafters to accent a country-style home, cast iron for a southern colonial, formal balustrades for a colonial or traditional home, finished oak planks for a craftsman-style cottage, exposed piping for a contemporary home; the possibilities are endless.

To maintain a sense of soaring height in the room below, treat the underside of the loft similarly to the room's ceiling. For instance, cover both with beaded board paneling, or use the same color paint for both. If the loft will rob light from the space

below it, add additional electrical lighting and/or more windows to make up for the loss below the loft. At the same time, plan your extension of electricity and supplemental lighting for the loft area.

Planning Rooms

While planning for any basement, attic, or garage conversion, be sure to include thoughts about telephone access, sound systems, and intercoms. Conveniences like these may save you lots of trips up and down stairs to answer the telephone or communicate with other family members. Wiring is easy to install, so plan to include telephone jacks and other convenience outlets in several locations, particularly if the conversion space is large, will be divided into two or more rooms, or will be used as a home office.

Noise Control. Also think about controlling noise. Thick wall-to-wall carpeting and a high quality pad absorb much of the sound that otherwise passes through attic floors. For additional sound control, place sound-deadening material below any underlayment or subfloor you install. You can also add fiberglass insulation between joist bays. This is about all you can do to reduce sound transmission without adding considerable expense to the project.

Give the same sound consideration to basement and garage conversions. Certainly, the noise you'd want to reduce for a basement, especially if you plan on bedrooms or an office in the space, would be that from upstairs. An easy way to accomplish noise reduction through floors and into basements is by the installation of thick insulation between the basement's ceiling joists. With garage conversions, it depends on what you have planned for the space. For a workshop, where the use of power tools could easily disturb others in the house, consider insulation and soundproofing board on the walls that separate the garage conversion

from home living spaces. For a recreation room that may be separated from existing bedrooms by only a 2x4 wall, you could install insulation in the existing wall, build a secondary wall 1 or 2 inches away from that existing wall, fill the new wall with insulation, and cover it with soundproofing board. The air space between the two walls will greatly assist in reducing noise transmission.

Room Dimensions. Building codes may vary from region to region with regard to ceiling heights and square footage for habitable rooms. Be certain to check with your local building department before starting any remodeling work.

According to most building codes, all habitable rooms must have at least 70 square feet of area with not less than 84 inches in each horizontal direction. For attic, basement, and garage conversions this standard is generally easy to meet, so from a practical standpoint the size of most rooms is governed primarily by the size of the furnishings to be used. A 70-square-foot bedroom, for example, is hardly a suitable location for a double bed. Keep in mind that the lack of abundant natural light in a basement can make rooms feel more

cramped than they might feel in an attic or garage conversion; don't assume that a comfortable small room upstairs will feel the same if you replicate it downstairs.

Building codes also normally require that living areas have a minimum ceiling height of 90 inches over at least one-half of the space. The only exceptions are bathrooms, kitchens, and hallways, which can have a ceiling height of 84 inches. If a quick measurement between the basement floor and the underside of the joists or the attic floor and underside of the ridge doesn't show the minimum headroom spacing, it may not be possible to obtain a building permit. Note that the headroom clearances are from the finished ceiling to the finished floor, so take all finish materials into account.

Design Considerations. Establishing the best goals for your remodeling starts with knowing your lifestyle. Be aware that not everyone is likely to have the same priorities. Get to know your family as an architect or interior designer would. Interview everyone and invite suggestions. You might like more privacy to rest and relax, for instance, while a teenager may crave a space large enough to invite noisy

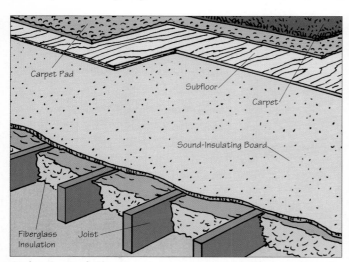

Noise Control. *The easiest way to gain maximum soundproofing is to place fiberglass insulation and soundboard beneath the subfloor of an attic or in the ceiling space of a basement. Wall-to-wall carpeting and a carpet pad complete the system.*

Room Sizes

The table below shows the minimum size of various rooms as set forth by the U.S. Department of Housing and Urban Development. For your planning, however, note that these government standards are bare minimums. An additional column, with more desirable minimums, is also provided.

Room	HUD Minimum	Preferred Minimum
Living room	11 x 16	12 x 18
Family room/Den	10½ x 10½	12 x 16
Great room	—	14 x 20
Kitchen*	—	—
Master bedroom	—	12 x 16
Other bedrooms	8 x 10	11 x 14
Bathrooms (full)	5 x 7	5 x 9

* The size of the kitchen will vary greatly with the selection of cabinets and the appliance layout.

▲ *Bright colors, plus the use of black and white, transform this basement playroom/ family room. The room is designed to accommodate many activities, including reading, entertainment, and children's play.*

▲ *A cramped kitchen benefits from relocating storage and expanding into the dining room. A skylight and shaft provide bountiful natural light and further open up the area.*

friends over. Start with shared objectives, then compromise where interests conflict.

Don't be restricted by the current functions of a room. Take a fresh approach and consider how a room can be reconfigured. The following pages contain general suggestions for rooms by function so you can see whether changes make sense.

Planning a Kitchen

Building or remodeling kitchens usually involves making them more efficient and perhaps expanding them into an adjacent area. It's costly to move an entire kitchen. In planning an expansion, consider the orientation. It's preferable to have the kitchen on the east or southeast to catch the morning sun and avoid afternoon sun. Kitchens generate their own heat and can become stifling with additional sun warmth. Your lifestyle counts in kitchen location, though. If your family leaves before sunrise, you may want to eschew conventional notions and have a western exposure to catch sunsets at day's end.

If you enjoy outdoor meals or cooking, you'll want outside access from the kitchen. A deck or porch at kitchen level directly outside is ideal for the inveterate outdoor chef. Make sure it's easy to maneuver from the kitchen to the deck with food and equipment.

Be sure to keep the dining room close to the kitchen and with easy access to it. Better yet, decide whether you really need a dining room in the first place. If casual is your lifestyle preference, consider expanding a cramped kitchen into an existing dining room and creating one cooking and eating space. Alternatives include making the dining room multipurpose, functioning as a den, home office, or guest bedroom as well as for dining.

Planning a Family Room/Playroom

Determine the purpose of the family room first, then plan its position. You'll avoid a room that's under-utilized because it doesn't suit your family's current real interests or needs.

A family room oriented to the south or west will have inviting natural light in the afternoon. It's practical to place the family room/playroom near the kitchen. You can use it for casual eating, and parents can keep an eye on children from the kitchen while cooking. Plan outside access from the family room if you often use the yard, deck, or patio.

Attics converted to playrooms or family rooms can be delightful. Removed from the rest of the house, they're especially suited to more restrained family room uses such as television watching, board games, quiet hobbies, and the like. Children generally like the treehouse feeling an attic space provides, especially when interesting angles from dormers and ceilings can encourage a sense of play. Attics aren't usually practical for noisy play, though.

The key to a great basement or garage playroom is versatility. Plan the space so that it can be used for a variety of activities. Wheeled storage cabinets, for example, can be rolled out of the way for large family gatherings and parties. Look for furniture that can be moved easily, build adaptable storage units, and install wall and floor surfaces that can withstand the hard use this room will probably sustain. Vinyl flooring, for instance, stands up to food spills, and area rugs can give it "warmth." Wood paneling is more likely to survive chance encounters with pool cue sticks than drywall is. An intense color or pattern can be overwhelming if used on all of the walls, but a spot of color (one brightly painted

wall, for example) or some diagonal paneling can do wonders for a room.

There are no particular electrical requirements for the average play-room, but again, the best plan is a versatile one. Extra cable outlets provide the opportunity to place a television in various locations, and several telephone jacks can be surprisingly convenient. Look closely at

▲ Relocating a powder room and storeroom made it possible to open up this kitchen to an adjoining family room.

▲ Maintaining a ventilated space above an attic conversion unifies the angles of this room designed for quiet activites.

your family's interests and plan for anything that might involve electricity, including lighted shelves for collectibles or outlets for exercise and fitness equipment.

Planning a Living Room

Many people want to have a living room, whether it's frequently used or not. If yours will function mostly as a reception area for guests, position it close to the front hall. Build in a means of closing the living room off from the rest of the house so that private areas are kept private. A separated living room ensures that you need not have the entire house neat and tidy when a visitor arrives.

Capitalize on an infrequently used "front parlor" living room by making it a dual-function room. When not entertaining use it as a quiet space for studying, reading, or with a desk tucked into the corner, a place for writing or other non-disruptive activities.

Depending on your lifestyle, you may want to forgo a formal living room and create an open, all-purpose communal great-room space. The area can function as a family room, home theater space, conversation center, and even dining room all in one open area. Keep in mind that it's hard to find privacy in such open plans.

Planning a Bedroom

Make sure any bedroom you plan is large enough to accommodate the size bed you want, along with any other furniture. The most important factor (and sometimes the most difficult) in bedroom planning is the provision of an emergency exit. Building codes generally require that every bedroom, including those in basements, have direct access to a window large enough for egress (at least 5 square feet) or an exterior

▲ Easily maintained wood paneling emphasizes the magnificent sweep of this family room in an attic space.

▲ Clever design, incorporating a built-in wall unit with video and a fireplace, transforms a basement into a general-purpose, communal family room.

▲ *Cabinetry makes sense of the odd angles of this gable-end bedroom. White maintains the sense of space.*

▲ *A study and play space with clouds seeming to rise to the ceiling makes use of a loft to double usable areas.*

door that can be used in an emergency (see "Building Codes and Basement Windows," page 132). In basements the door can't lead to a bulkhead door.

Privacy is another important consideration for bedrooms. All bedrooms should be removed from the activity rooms commonly shared. Create sound buffers by placing closets, bathrooms, and storage rooms between bedrooms and activity rooms. Parents also may want privacy from a child's bedroom, especially if the child's room also functions as playroom.

Master bedrooms can successfully incorporate other functions in addition to sleeping, such as an exercise area, work or study area, video-viewing area, or comfortable seating area. Individual preference will dictate how close you want to be to children's bedrooms or how much you want to make the bedroom a master suite in the full sense. Optimum positioning places bedrooms

on the eastern side or southeast to take advantage of the morning sun, a psychological boost when starting the day.

Closets. The design and size of closets in a bedroom depend in part on who is to use the room. A

modestly sized closet will probably suffice in a guest bedroom, but a master bedroom calls for an extra-large closet. Manufacturers of closet shelving and storage systems are good sources for closet-design information. Such systems allow you to pack the most storage into the least amount of space. Consider building at least one cedar-lined closet to help keep moths away from clothing.

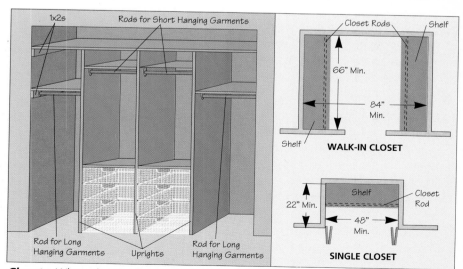

Closets. When planning for closets in your conversion, take stock of the items that you have and allow for long and short hanging garments. Install shelves and drawers to accommodate other clothing or things you want stored in the closet. These are the standard sizes of closets typically found in bedrooms. Note that the walk-in closet is a minimum 84 in. wide and the single closet is a minimum 48 in. wide.

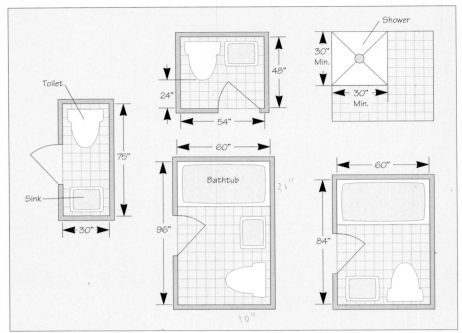

▲ *A pocket door makes this bathroom, complete with a shower and whirlpool tub, fit into the attic. A skylight and circle top window flood the space with light.*

Planning a Bathroom. *These are minimum dimensions for bathrooms. Spaced any closer together, the facilities would be difficult to use comfortably.*

Planning a Bathroom

In most cases the location of a bathroom is determined by the accessibility of drain, waste, and vent stacks. Plumbing is easier to install and less expensive if it can be tied into existing drain and vent pipes. Keep costs down by locating the bathroom either directly above, below, or back-to-back with the plumbing of the kitchen or another bath. Usually, the toilet most complicates the installation of a new bathroom because it requires a bigger drain line than does a sink or shower.

According to building codes, the headroom in a bathroom can be as low as 84 inches—6 inches lower than the standard for other rooms. But note that the most common criticisms of bathrooms are that they are too small and that there aren't enough of them. One possibility is to add a master bath while leaving the old bathroom for general use. Plan a master bathroom that can conveniently accommodate two adults at the same time.

Master Baths. Today, the size of the master bathroom is equally important as that of the main family bathroom. If you can do so, commandeer a small bedroom, dressing room, or storage room to create a large master bath. The larger size will enable you to install both a shower and a whirlpool, as well as double sinks and ample vanity space. A luxurious bath will entirely change the ambience of the master suite.

Design for Privacy. In all cases, privacy is a prime consideration in bathroom design. Ideally, bathrooms should be located so that they don't open directly onto daytime activity areas of the home. For instance, place bathrooms off a hallway rather than opening directly into a room like the living or family room. All bedrooms should have a bathroom that's easily accessible, either directly or just down the hall.

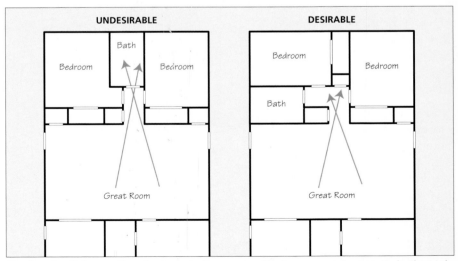

Design for Privacy. *Plan the layout of rooms and hallways so entrances to baths and bedrooms are shielded from the direct view of areas where you'll entertain.*

Planning a Storage/Utility Area

Whether it's in a section of the garage or basement or on the main floor, you'll need room for storage and utilities. In most cases, a total reorganization can do wonders in making storage spaces work most efficiently.

In cold climates place these rooms on the north side of the house as a buffer to cold. If summer heat is more of a problem, place them on the south side. Because laundry centers create their own heat, place them on the north as well. Position efficient storage at various nearby locations throughout the house for convenient day-to-day use. On the other hand, use the strategy of storing off-season clothes and accessories in a remote storage area to free up frequently used closets. In any case, don't assume that you can rob storage space in remodeling and not replace it. You're better off planning on additional storage right from the beginning.

Planning a Workshop

The solid floor and sturdy walls of a basement or garage lend those areas to the wear and tear that's typical of most home workshops. Because the height of a basement may be limited, additional horizontal space may be necessary to maneuver materials back and forth. Provide plenty of electrical outlets on one or more dedicated 20-amp circuits. Depending on the type of equipment to be used in the workshop, both 220-volt and 110-volt outlets may be necessary. Given the proximity of furnaces and water heaters—and often the lack of ventilation—in basements and garages, plan to conduct staining, painting, and other tasks that involve flammable vapors in other more openly ventilated locations.

Shop Lighting. Proper lighting is critical in workshops, so don't skimp on it. The ideal combination employs fluorescent lights for general lighting and incandescent lights for task and supplemental lighting. Wire the incandescents to the most-used wall switch, since they hold up better to frequent use, and wire fluorescents to another switch so they can be turned on when you plan to be in the workshop for longer periods of time. Fluorescent fixtures hung from short lengths of lightweight chain are easy to move. On the other hand, those designed to fit inside floor-joist or rafter spaces will fit flush against the ceiling to provide more headroom. Look for special shatter-resistant fluorescent and incandescent bulbs.

Dust Control. Keeping sawdust contained is a most important precaution. Not only can dust be a nuisance when it ends up in living areas, it can also be damaging to mechanical equipment. Furnaces call for special care. Dust quickly clogs the filters on forced-air furnaces, and oil furnaces are adversely affected by excessive dust. More importantly, excessive dust accumulated around combustion appliances like gas- or oil-fired water heaters poses a significant fire hazard. Under the right circumstances excessive dust could even cause an explosion.

Isolating the shop from adjacent rooms with partition walls is the best dust-control strategy. Help to keep dust confined to the workshop by outfitting each door that enters the space with weatherstripping. A portable dust-collection system is a must for confined workshops, especially those in basements. Hoses from the dust-collection system connected to each woodworking machine will cut down on dust pollution considerably but will not eliminate dust entirely. At the very least, connect a shop vacuum to each source of sawdust as work is under way.

Planning a Home Office

A home office is likely to be filled with electronic equipment that includes computers, printers, photocopiers, fax machines, and so on.

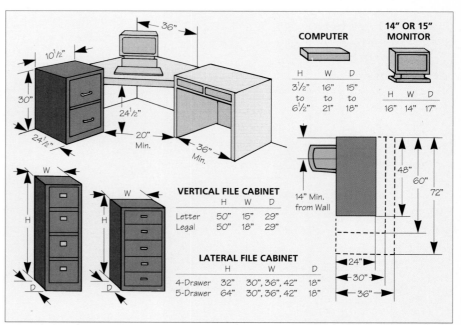

Planning a Home Office. *A new attic, basement, or garage conversion may make an excellent home office. Be certain you provide enough space for desks, file cabinets, and other pieces. Also, make sure you install enough electrical outlets for your equipment.*

▲ *Home offices typically have electronic equipment like computers and printers, so be sure you have enough receptacles on at least two circuits.*

Allow for plenty of electrical outlets, and as a precaution, divide them into at least two separate circuits if possible. Some pieces of home office equipment, such as laser printers, have significant power requirements, and if a circuit breaker trips while you're using the computer, you may lose important data.

Decide whether more than one telephone line is necessary for your home office. It's possible for a computer modem and a telephone to "share" one line. But if you frequently use the computer for on-line research, it may be more convenient to put the telephone on a second line. Convenience and organized record-keeping are other reasons to have two telephone lines. If one line is an extension of your home number and the other is for business only, you can answer personal calls without leaving the office and track business-call expenses separately.

Although it depends on the kind of work you'll do, most offices should have bookshelves, as well as storage for files, office supplies, and the like. Make every square foot count. If a tall four-drawer file cabinet seems too awkward for the room, for example, consider using a pair of two-drawer units with a piece of plywood on top. This arrangement allows for plenty of file space and serves as a stand for a printer, photocopier, or fax machine at the same time. Here are other factors to consider:

▶ Attics. An attic is a good choice if you do not need a separate entrance for business traffic.

▶ Late working hours. Will you work late at night? Will night work disrupt the sleep of other family members? For instance, don't combine a home office with a child's bedroom in the attic. Or if you set aside part of a master bedroom as an office, be sure late-night work won't disturb your spouse.

▶ Home office tax deductions. Have you checked that you can legitimately declare the remodeled space as office space? Does your design conform to the regulations

▲ *A treetop office with handsome paneling and a built-in desk is efficiently tucked into an 8x9-foot space.*

▲ *A quiet office in an attic conversion melds modern lines with traditional furnishings, playing off the angles.*

for home office space? Is it completely separated from other areas, for example, so 100 percent of the space is work related?

▶ Insurance. What liabilities will you incur if business guests must go through your living quarters to reach the office area? What about insuring your office equipment?

▶ Zoning ordinances. Are there zoning restrictions that may influence your location of a home office?

▶ Work at home versus occasional use. What is the purpose of the office now? How often will it be used? Do you need office space for two? Can the office be shared efficiently? Should you plan on using the home office full-time at some future date and incorporate anticipated needs with current design?

Fitting in the Furniture

To see whether a room's new function will really work, fill in your redesigns with the furniture as you did in your old plan to help you establish the minimum proportions needed. In a master bedroom, for instance, you probably want wall space to accommodate a bed plus two end tables and a television positioned across from the bed a convenient viewing distance. Use bedroom closets as sound barriers to ensure privacy as well as providing storage. Make sure you have convenient closet access by not placing furniture directly in front of closet doors.

Also, consider how the room will be used. Plan seating in a home theater, for example, so that someone pass-

Clearance Requirements

Following are some general furniture clearances. Use them to determine whether a room lends itself to a workable layout for its intended use.

■ Doorways must allow for clearances for doors and access. You need 36 inches of free space into a room; outside entrances need 48 inches. Clearances help determine which side of patio or sliding doors should be fixed and which should open.

■ Drawer storage cabinets call for 40 inches of clearance.

■ Closets need 36 inches of minimum clearance.

■ To allow enough room to make them up, beds require at least 22 inches of clearance.

■ Desk or worktable seating needs 36 inches of clearance.

■ Dining areas need at least 32 inches of clearance from the table to a wall for someone to rise from a chair. Allow 36 inches for someone to walk behind a seated person.

■ Sofas and chairs require a minimum of 18 inches of clearance in front of them. Use this clearance between a sofa and coffee table, or to be sure that a seated person's feet will not jut into a traffic lane.

■ Main traffic lanes should be at least 48 inches wide: minor traffic lanes at least 24 inches wide. Traffic lanes are the main ingredients to determine whether your furniture placement and room details will work. Make your traffic lanes as close to the most natural walkways as possible. Natural patterns often are direct lines from doorway to doorway or from the doorway to the main seating area. If the distance across a major seating area is too long, break up seating into more than one grouping.

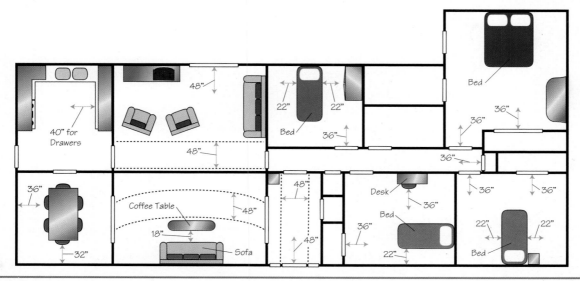

Clearance Requirements. *Maintain the minimum clearances recommended here for the most comfortable use of doorways, furniture, and traffic areas.*

ing through the room won't have to walk between viewers and the video screen. Check that details such as electrical outlets and light switches are conveniently positioned if you've relocated doors. The electrical utilities should also relate to your prospective furniture arrangements. If you're planning a media wall with electrical demands, for instance, make sure the outlets are positioned to service it.

Door placement, wall lengths, and furniture arrangements all work together to make the most of space. It's frustrating to have to discard a great furniture arrangement because a wall is just 6 inches too short when the doorway easily could have been positioned farther away. While these clearance recommendations are not carved in stone, they are guidelines that make design efficient and convenient. Use the clearance recommendations in the box to discover flaws in your plans and eliminate those that the furniture placements reveal.

Lighting

General Lighting Requirements

To provide a suitable amount of natural light, building codes generally require that all habitable rooms have an amount of glazing (window glass area) equal to 8 percent or more of the floor area. For daylighting, it doesn't matter whether the glazing is fixed or operable. This amount of daylighting is difficult to achieve for rooms located partially below grade, so most building codes allow an exception to the requirement for these areas.

Natural lighting can be entirely forgone if artificial lighting provides an average of 6 lumens per square foot over the area of the room. Lumens

are a measure of the total amount of light emitted by a light bulb; the more light a bulb produces, the higher its lumen rating. Six lumens per square foot is not difficult to achieve. To determine your needs, measure the length and width of the area in question, then use the chart to see which kind of light is best and how many bulbs are needed to meet the code. Once you've selected the kind of bulb you want, it will be easy to determine the kind of light fixture needed.

Note that the 6-lumen figure is for general, or ambient, lighting and is an average requirement for each room. The provision for general lighting is the first priority for planning room lighting. After that, task lighting and accent lighting can be added as desired. Special-purpose rooms like bathrooms and home offices may have additional lighting needs.

Some parts of the country are prone to frequent, though short, power outages—particularly during stormy seasons. It's a good idea to plug an emergency light into one or two outlets in basement and attic living spaces, though they're not required by building codes. Small inexpensive emergency lights are powered by a rechargeable battery that maintains a charge while plugged into an electrical outlet. When household power stops, the battery takes over to provide enough light for occupants to navigate through otherwise dark rooms.

Designing with Light

To maximize the effectiveness of lighting, use light-colored surfaces wherever possible; this helps to reflect light around rooms. Dark paneling or carpeting, on the other hand, tends to "soak up" light. Use a variety of light sources, if possible, to provide maximum flexibility when it comes to setting a mood or producing extra light for activities.

Light Quality. Light quality is worth ample consideration, especially for basements, which rely heavily on artificial light. Even if the quantity of light is adequate, the quality of the light can make or break a room.

Lighting quality can be generally described in terms of the "coolness" or "warmth" of its color. This lighting temperature is measured using the Kelvin Scale, abbreviated K. Cool light emphasizes blue and green hues while warm light plays up yellows and reds. The color of the light you get depends on the kind of

▲ False eyebrow windows at basement window height give this room the appearance of being above ground, as do the light colors used throughout.

light bulb you use. Conventional fluorescent bulbs produce cool light while incandescent halogen, flood, and standard bulbs emit warm light. Because people tend to prefer warm light, they have traditionally avoided fluorescent lighting in homes. Modern fluorescent bulbs, however, are available in varieties that closely approximate the warmth of incandescent lighting. Add this quality to their energy efficiency and the new fluorescents become well worth consideration. If you prefer warm light, look for fluorescent bulbs rated at less than 3000 K.

Lighting in Basement Conversions. Lodged beneath an existing floor system and kept from daylight by tons of earth, most basements present a real challenge for lighting design. Unlike above-grade attic and garage conversions, basements can't be easily brightened with the addition of a window or skylight. In many cases, the only basement windows to work with are those mounted high in walls. Another complication relates to space proportions and dimensions. With so much floor space beneath relatively low ceilings, basement rooms can feel confining and uncomfortably cave-like.

Since you can't count on supplementary natural light, provide enough light to make the basement functional as well as attractive. You need ambient, overall lighting. In addition, you need task lighting, which puts a high level of illumination on the surfaces where you need it. In addition, make sure that there's adequate lighting for game tables and reading if these activities will take place in the basement. A basement's low light level makes it an ideal location for a media room or home entertainment center. You can totally control light levels and reduce glare without interference from the outdoors, as in above-ground installations. Also, low light levels don't create a problem for activities normally undertaken at

Types of Fixtures

Not many people install chandeliers in their basements or attics, but most other types of fixtures may be used. Pay attention to headroom when choosing ceiling fixtures. Also, different fixtures distribute light in different ways: Some illuminate broad areas whereas others spotlight small spaces. Be sure to choose the fixture that best suits your needs.

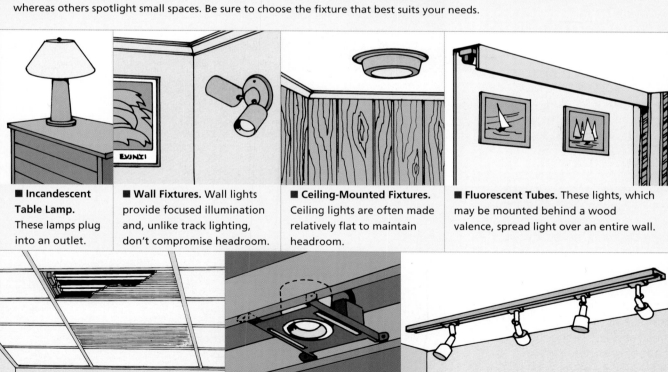

■ **Incandescent Table Lamp.** These lamps plug into an outlet.

■ **Wall Fixtures.** Wall lights provide focused illumination and, unlike track lighting, don't compromise headroom.

■ **Ceiling-Mounted Fixtures.** Ceiling lights are often made relatively flat to maintain headroom.

■ **Fluorescent Tubes.** These lights, which may be mounted behind a wood valence, spread light over an entire wall.

■ **Fluorescent Fixtures.** Many fluorescent-tube fixtures are designed to fit into a suspended ceiling system or between ceiling joists.

■ **Can-Type Lights.** These lights are recessed into joist cavities to provide unobtrusive spot or floor lighting.

■ **Track Lighting.** This kind of lighting can be fitted with floodlights for general illumination or with spotlights for accent lighting.

night, when all the windows in the world won't make a difference.

Plan atmospheric lighting to make the space attractive. Place light in the corners of a basement to banish the shadows; cast a glancing light across an interesting wall to bring out its texture or to spotlight a favorite object such as a sports trophy. No matter what the ultimate function, you'll want to have some welcoming lighting in the basement. Use dimmers to set the lights low, but be sure that general light levels can be bright enough on entering the basement to keep the area from being gloomy.

Position switches so that you can exit the basement well lit, turning off lights when you're on a good footing above. Install emergency lights near your service panel and along a path toward it, especially if you've remodeled so that the panel isn't conspicuous. Avoid the mistake of putting in only one kind of lighting, such as light panels over a dropped ceiling. A variety of lighting sources will more closely simulate the natural lighting experienced when the sun floods the windows or when individual lamps are turned on upstairs at the day's end.

Lighting in Attic and Garage Conversions. Attics and garages lend themselves to natural lighting because you can easily add windows and skylights. As a general rule for natural lighting, figure you'll want a total glass area of at least 10 to 15 percent of the room's floor area. Place the skylights where they provide efficient light. For example, spread skylights along the length of a roof or cluster some in a dark area that's not illuminated by conventional windows at either end of the attic.

Light tubes are other options to consider. Generally, these are constructed of a fixed exterior skylight coupled to a flexible reflecting tube that carries light down to a translucent glass-covered opening in the ceiling of the room below. Use them when you want diffused light rather than a view. They provide a solution when you don't have a clear path between the inside ceiling and the roof. Light shafts, especially when they widen inside the room, provide natural daylight in areas where it otherwise would be impossible to position a skylight or roof window. They also provide privacy as well as light.

Although skylights and windows will allow plenty of sunlight to spill into attic spaces during the day, low ceilings may pose problems for lighting designs. Consider recessed ceiling fixtures that mount flush with ceilings.

Garage ceilings are normally at least 96 inches from the floor. You should be able to mount almost any kind of ceiling light fixture in garage conversions, including fan-light combinations, with little worry of hanging them too low.

▲ *Skylights provide plenty of natural light for daytime illumination of this attic conversion.*

▲ *The flow of the kitchen is smooth and effortless into the dining room, which was converted from an adjacent garage.*

Sizing Up the Project

2

B efore you spend any money on a remodeling project, it's important to size up the situation. Sometimes you can use existing structural elements to your advantage without making a lot of expensive changes. Take a close look at the attic, basement, or garage to determine what can be done. Then learn enough about building codes to keep your project on the right course. It's much harder to correct mistakes than to avoid them in the first place.

Surveying the Attic

Once you decide to convert an attic to living space, you must determine whether the job is possible. Not every attic can be converted to living space, and some of those that can be converted aren't worth the effort. To find out more about the possibilities for your attic, head up there with a tape measure and flashlight.

Checking for Trusses

A roof that was built using trusses, rather than individual rafters, can't be converted. If the attic is filled with diagonal framing members, called webs, it was built with trusses. Webs give each truss its strength and can't be removed without causing the trusses to fail. If the trusses fail, the entire roof has to be rebuilt. If this is the situation with your roof, but you still need more space, your best alternative is to convert the basement or garage instead.

Investigating the Attic

If the roof is supported by standard rafters, there may be enough room to convert the attic to living space. Before you can know for sure, however, you must investigate further. Temporarily nail some 1x4 or 1x6 boards (or plywood if there's enough room to get it up there) across the tops of the attic floor joists. The boards prevent you from accidentally stepping through the ceiling and provide a safe platform from which to work. Never step directly on insulation; it's almost always supported by drywall alone and can't support your weight.

Headroom. According to code, there has to be enough headroom in an attic conversion to enable you to move about comfortably and safely without clobbering your forehead whenever you turn around. To get an idea of whether your attic is even close to meeting this standard, take a quick measurement between the ridge and the top of a floor joist. If the distance is not at least 91½ inches (1½ inches accounts for the thickness of a finished ceiling and floor), you can't convert the attic without major work and a variance from local building officials.

The complicating factor in figuring attic headroom is that the ceiling slopes. Because of this, some of the floor area is worthless for use as walking space even though it may be perfectly suitable as storage or working space. Recognizing this, the building code calculates headroom requirements in the following way: To begin with, all living space in the attic must be at least 70 square feet in size and measure no less than 84 inches in every horizontal direction. A small room, or one that's long and narrow, is not considered suitable living space. In addition, at least 50 percent of the floor space must have at least 90 inches of headroom. The rest can have as little as 60 inches. Finally, portions of the room with less than 60 inches of headroom are not considered living

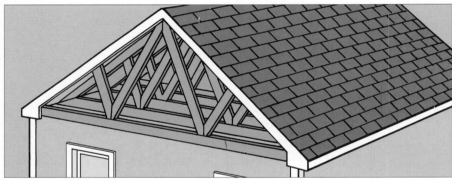

Checking for Trusses. *If this is what you see inside your attic, forget about an attic conversion. Trusses can't be removed or altered.*

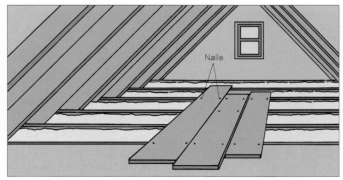

Investigating the Attic. *Lay boards or plywood over the tops of exposed joists before spending time in the attic. Secure the boards temporarily so they don't shift as you walk on them.*

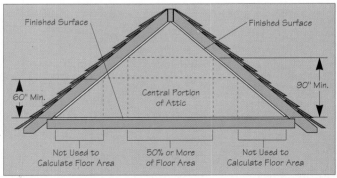

Finished Surface

Finished Surface

60" Min.

90" Min.

Central Portion of Attic

Not Used to Calculate Floor Area

50% or More of Floor Area

Not Used to Calculate Floor Area

Headroom. *This diagram shows how to determine whether the attic has enough headroom for a conversion to living space.*

space, so you should not count them in the calculations above.

If the roof framing has collar ties, take headroom measurements from the floor to the underside of the ties. Also, if the flooring doesn't exist yet, subtract 1½ inches from each measurement you make to account for the thickness of floor and ceiling finishes.

Ventilation. During the warm seasons an attic becomes the hottest area of the house. The problem of overheating must be addressed before turning the attic into a living space. Proper ventilation through the roof spaces, as well as though the room itself, is crucial for comfort even if you plan to air-condition the space. By addressing the ventilation possibilities now, you can determine whether converting the attic is worth the trouble.

You need at least one window on each end of the attic to encourage an adequate flow of air. You may wish to add at least one dormer window or a ventilating skylight to improve ventilation, even if the dormer isn't needed to improve headroom. If a chimney at one end of the attic prevents you from installing a window there, a dormer or skylight nearby can provide the necessary air circulation.

Measure the depth of the existing rafters to see whether there's enough room for fiberglass insulation plus a 2-inch air space above the insulation. The air space ensures that moisture migrating into the insulation is carried away. It also helps to keep the roof assembly cool in warm weather by providing a ventilation path immediately beneath the roof sheathing. The ideal ventilation pattern draws in the air through soffit vents and exhausts it through a ridge vent; however, any combination of vents placed low and high encourages air movement.

Rafters. Rafters that sag noticeably may be dangerously undersized. Sags may be the result of too many layers of roofing or improperly sized rafters. To determine the severity of the sag, stretch a string along one rafter from the bottom edge at the top of the rafter to the bottom edge at the lowest point you can reach. Measure the amount of sag at the midpoint of the string. If a group of rafters sags more than ½ inch, you may have to install a structural kneewall to support them.

CAUTION: It's imperative that this structural kind of work be done properly, so consult a structural engineer before proceeding.

If only one or two rafters sag, check them for cracks, open knotholes, or

other damage. It's usually easier to repair damage than to replace the rafter. Straighten the rafter if possible, and bolt new wood over the damage, much as you would splint a broken arm. Sometimes you can use a 2x4 to straighten a rafter. If this quick fix doesn't work, the rafter may have to be jacked into place. Rafter jacking is a job for a professional.

Floor Framing. Check the attic floor joists for damage and sag, using the same technique used for checking rafters. Usually, sags result from insufficient support rather than undersized joists. If joists overlap near the center of the house, they must be supported by a wall or a beam directly underneath. But if part of that support has been removed, as is sometimes done when two small rooms are combined into one large one, the joists above can sag.

Structural support must continue all the way to the foundation, so check the basement or crawl space to see whether support is missing there. If the house was built correctly, you'll find a beam or another wall directly beneath the support wall. If you don't see one, a builder or an engineer can determine the proper combination of beams and/or posts required for proper support.

Water Leakage. Inspect the underside of the sheathing and the sides of the rafters for brownish stains that may indicate a leak in the roof. You can rarely repair leaks from inside the attic, so count on making the repairs from outside. If you find a stain, be sure to investigate further; some stains may be due to an old leak that has since been repaired. If the area feels spongy when you probe it with a flat-blade screwdriver, the leak is active and must be repaired. Another way to check for leaks is to visit the attic during or just after a hard rain.

Leaks are commonly found around flashing, so check around all places

Ventilation. *Measure the depth of the existing rafters to see whether there's enough room for fiberglass insulation plus a 2-in. air space above the insulation.*

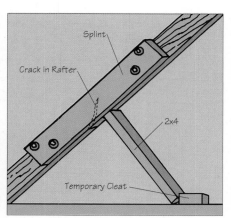

Rafters. *Push a damaged rafter back into place and use a splint of lumber the same dimension as the rafter to reinforce it.*

where the roof has been penetrated, such as plumbing vents and chimneys. Replace or repair all suspect flashing.

Insect Problems. Although attic lumber isn't infested as easily as wood that's closer to the ground, keep an eye out for signs of powderpost beetles, carpenter ants, and non-subterranean termites. Look for swarming insects, a series of pinholes in the wood, and small powdery piles of sawdust beneath affected wood. If you suspect an infestation, rap the wood with your knuckles or the handle of a screwdriver. Infested wood makes a sound different from that of solid wood. Have infested areas treated by a professional exterminator before work begins.

Chimney Problems. It's acceptable to have an airtight chimney within a living space, as long as you have it inspected by the local fire department. A building inspector can advise you on the relevant building codes. There's not much you can do about moving a chimney. Take some measurements to determine its size and shape, then figure out how it fits into your plan. Measure each side of the chimney to determine the amount of headroom you'll have when walking around it. Then examine the chimney for loose mortar or cracks in the masonry. Cracks are a particular hazard if the flue doesn't have a fireproof lining. Joists, rafters, or other framing that's closer than 2 inches to the chimney at any point is a fire hazard and must be corrected immediately. Make sure all combustible materials are kept well away. Unfaced fiberglass is not combustible. If the chimney is made of metal, it must be enclosed. Check local codes for permissible solutions.

Surveying the Basement

Once you have an idea of how you want to remodel the basement, a bit of detective work is necessary. Uncovering and solving potential problems at the start means being faced with fewer surprises and less expense later.

Not every basement can be converted into living space, and not every one that can is worth the effort. If, for example, the basement is short on headroom, the solution (lowering the floor level) involves more effort and expense than it's worth. Likewise, if water problems can't be eliminated despite your best efforts, you can't turn the basement into a comfortable and healthy living space. Spend some time getting to know your basement before jumping into a remodeling job.

Types of Basement Walls

The kind and condition of the walls found in the basement has a lot to do with how easy or hard the basement will be to remodel. Basement walls, of course, are the inside surfaces of the foundation. They can be made of concrete block, poured concrete, stone, or pressure-treated wood. Though some are easier to work with than others, none of the foundation types automatically prevents you from remodeling the basement. It's easy, for example, to install drywall or paneling on the walls of a pressure-treated wood foundation. The procedure is the same for installing drywall on wood-framed walls. A stone foundation, on the other hand, sometimes has water problems that are difficult to remedy due to the irregular nature of the materials. Walls of concrete block or poured concrete are the most common.

Concrete Block Walls. A foundation made with concrete blocks is easy to

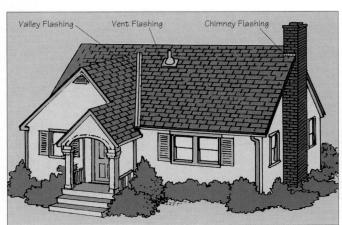

Water Leakage. *From the roof, check all flashing for signs of deterioration and replace faulty pieces. Also, check the underside of the roof sheathing for signs of leaks in the shingles.*

Chimney Problems. *Look for cracks in the masonry and mortar. As long as combustibles are kept away, an airtight, crack-free chimney is safe and can add interest to the decor.*

identify because of the grid pattern created by horizontal and vertical mortar joints. Each block has a hollow core, and the inside and outside faces of the block are connected by integral webs. The hollow structure of the block makes it lightweight and easy to work with and allows the wall to be strengthened by mortaring reinforcing bar, or rebar, into the cores. Most blocks are identified by their nominal dimension, which is the measurement used to calculate how many blocks are needed in a wall. Nominal 8x8x16-inch blocks, which are typically used in residential construction, actually measure $7^5/_8$x$7^5/_8$x$15^5/_8$ inches. The extra $3/_8$ inch allows for the mortar joints.

Blocks, which are called concrete masonry units, or CMUs, in the building trades, are stacked one atop the other. Mortar placed between each row and each block bonds the units and results in a strong, solid wall. Because of this construction method, the ultimate strength and water-resistance of the basement wall depends on both the condition of the blocks and the condition of the mortar.

Poured Concrete Walls. A poured concrete wall is monolithic and has a smooth surface. To build such a wall, concrete (a mixture of sand, gravel, water, and portland cement) is poured into form work, usually made from steel or some other metal. Steel reinforcing bars are placed in the forms prior to the pour. The bars strengthen a concrete wall and help resist cracking.

Other Types of Walls. In some areas of the country, particularly the Midwest, builders may frame a house on top of a foundation of 2x8 or larger studs and plates that have been treated with chemicals under pressure to resist decay. This is a relatively new type of foundation. Sheathed on the outside with pressure-treated plywood and detailed carefully to eliminate water infiltration, the foundation can be insulated and finished like a standard framed wall.

Stone foundations can still be found in certain areas of the country, such as the Northeast, where some houses predate the availability of concrete. Although the kind of stone varies

according to that which was available locally, most of the foundations were laid up with mortar. To find out whether the foundation is in good condition, it's well worth having a mason inspect it before remodeling the basement.

Inspecting the Basement

Before beginning work, there's more you must know about your basement. It's easier to deal with a tricky basement problem before a small mountain of building materials is delivered to the front yard rather than after.

Foundation Cracks. Figuring out what to do about foundation cracks is more art than science. Hairline cracks in a concrete wall are sometimes the fault of improper curing; larger cracks are usually due to settling. Both kinds can be repaired with hydraulic cement if the crack isn't an active one, that is, if whatever caused the crack in the first place is no longer an existing problem. If the foundation is in the process of settling, however, or if

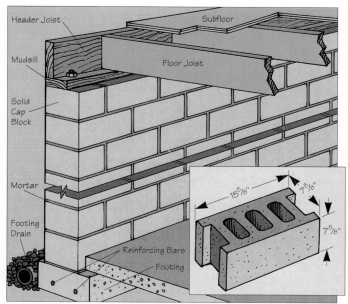

Concrete Block Walls. *This system consists of individual blocks bonded together with mortar. The blocks shown here are those most commonly used.*

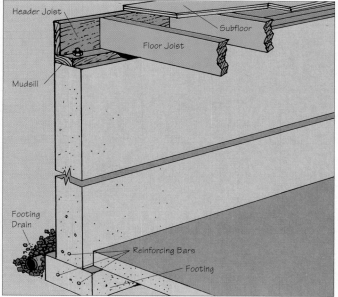

Poured Concrete Walls. *This kind of wall is poured from footing to top in one step. Steel reinforcing bars are added for additional strength.*

some other factor is stressing the foundation, cracks you patch today may open again tomorrow.

Inadequate Headroom. According to most building codes, a room in the basement must have a minimum ceiling height of 90 inches over at least one-half of the room. The only exceptions are bathrooms, kitchens, and hallways, which are allowed a ceiling height of 84 inches. Minimum headroom measurements are taken from finished surfaces. If a measurement between the basement floor and the underside of the joists doesn't meet these standards, it may be impossible to get a building permit.

Poor Access. Getting into the basement usually is not a problem. The stairs might have to be repaired, but at least they already exist. When it comes to basement bedrooms, however, having an exit in case of an emergency also becomes an issue. According to code, all bedrooms in the basement must have a means of emergency exit. A door that leads directly to the outside from a bedroom (and not to a bulkhead door) qualifies as an emergency exit. If no such door exists, there must be a window that can be used instead.

The requirement for such a window (called an egress window) is 5 square feet of operable area. If remodeling plans include a bedroom, make the egress issue the first order of business.

Lack of Air Circulation. All rooms become uncomfortably stuffy when the air in the room is not periodically renewed with fresh air. Building codes call for operable windows equal in size to at least 4 percent of the room's floor space. Don't confuse this figure with the amount of glazing needed to provide natural daylight in a basement. In a basement, however, this is not an easy percentage to achieve, so the code offers the following exception: If a room is served by an "approved" mechanical ventilation system that's capable of changing the air every 30 minutes, operable windows are not required. An approval must be given by your local code officials, however, so check with them before deciding to pursue this exemption.

Moisture Problems. Of all the possible roadblocks to making the basement livable, moisture problems can be the toughest to hurdle. Water is incredibly persistent, and under some circumstances can make its way

through walls that are considered impermeable. Another source of moisture is the condensation that forms as warm moist air reaches the cold surface of a masonry wall. Some moisture problems can be remedied, but major problems may call for professional assistance and considerable expense.

As an easy test for water problems, tape pieces of aluminum foil to various places on the walls and floor. Seal the edges tightly and leave the test patches in place for several days. If moisture droplets appear beneath the foil after several days, moisture is migrating through the masonry; if they appear on top of the foil, the problem is condensation from basement humidity.

When looking for water problems, investigate the underside of the first-floor subflooring for signs of leaks. Now is the time to fix faulty pipes and fixtures. Inspect the subfloor and the sides of the joists for brownish stains that may indicate an active leak or an old leak that has since been repaired. If the stain is spongy when you probe it with a flat-bladed screwdriver, an active leak exists somewhere.

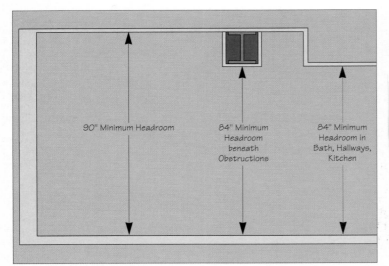

Inadequate Headroom. *You must have at least 90 in. over one-half the area of a basement room, but building codes allow 6 in. less under beams and in bathrooms, kitchens, and hallways.*

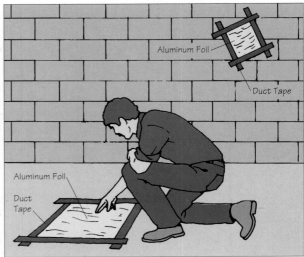

Moisture Problems. *Tape aluminum foil to sections of the basement or garage floor and foundation walls. If moisture collects (a) underneath, a seepage problem exists and must be corrected; (b) on top, a humidity problem exists.*

Insect Problems. The floor system of most houses rests on wood "plates" that are bolted to the foundation. If the house has a problem with wood-destroying insects, this is where you'll find the evidence. Check the outer 12 inches or so of the floor joists, the inside surface of the rim and header joists, and the wood frame of every basement window. Keep an eye out for signs of powder-post beetles, carpenter ants, and non-subterranean termites. Signs of insect problems include swarming insects, a series of pinholes in the wood, and small powdery piles of sawdust beneath affected wood. If things look suspicious, rap the wood with your knuckles; infested wood sounds different from solid wood. To search for rot or insect damage, use the tip of a screwdriver or scratch awl to poke at the rim and header joists, the plate, the ends of the joists, and window framing, even if the wood looks sound. Rotten or insect-infested wood yields easily. Building inspectors have been known to use a ski pole with a sharp-

Insect Problems. *Inspect the joists and other wood in the basement for dry rot and insects. Use an awl to penetrate the outer joist areas.*

Discovering Potential Health Hazards

Radon. Radon is a colorless, odorless radioactive gas that comes from the natural breakdown of uranium in soil, rock, and water. When breathed into the body, molecules of radon lodge in the lungs and lead to an increased risk of lung cancer. The incidence of radon is not restricted to certain areas of the country. Radon typically moves up through the ground and into a house through cracks and holes in the foundation, though they are not the only source. Because radon tends to concentrate in rooms closest to the ground, it's particularly important to test for the gas before converting a basement or garage to living space. If test results indicate that there's a problem, radon-reduction techniques are relatively easy to incorporate into remodeling plans.

How to Test for Radon. It's fairly easy to test a house for radon. Don't rely on tests performed on other houses in the area. Even homes that are next door to each other can have different levels of the gas. There are two basic types of radon tests:

■ Active tests call for special equipment and are generally the most accurate but require a specially trained technician.

■ Passive tests include a variety of inexpensive testing products available at hardware stores and home-center stores or by mail from state-certified testing laboratories. You expose these devices to the air inside the basement for a specified length of time, then mail them to a testing laboratory. Long-term passive tests offer a good indication of the year-round average radon exposure but must be in place for at least 90 days. Short-term tests, though not as accurate, can be completed in as little as 48 hours. Short-term testing is typically done under closed-house conditions, meaning you keep all windows and doors closed, except for normal entry and exit, and refrain from using fans or other machines that bring outside air into the house. You may operate the home's heating system normally while the test is being performed, but you can use the cooling system only if it doesn't draw outside air into the house. A short-term test indicates closely enough whether a major problem exists. If such a test indicates that there's a problem, it is recommended that another test be conducted to confirm the diagnosis.

If testing indicates a level of radon of more than 4 picocuries per liter (pc/l) of air (a standard measurement of radon), take steps to reduce the radon by employing a process called mitigation. A level of less than 4 pc/l (1.3 pc/l is considered average) is generally not worth the expense of mitigation.

How to Reduce Radon. Sealing cracks and other openings in the foundation is a basic part of most radon-reduction approaches. The U.S. Environmental Protection Agency (EPA) doesn't recommend sealing alone, however, because it hasn't been proved effective. In most cases, reduction systems that incorporate pipes and fans to vent air to the outdoors are preferred. Contact a licensed mitigation specialist.

Asbestos. Asbestos is a fibrous mineral found in rocks and soil throughout the world. Alone or in combination with other materials, asbestos was once fashioned into a variety of building materials because it is a strong, durable fire retardant and an efficient insulator. Unfortunately, it's also a carcinogen. Once inhaled, asbestos fibers lodge in the lungs. Because the material is so durable, it remains in the lung tissue and becomes concentrated as repeated exposure occurs over time. Asbestos can cause cancer of the lungs and stomach among those who have prolonged work-related exposure to it. Home health risks arise when age, accidental damage, normal cleaning, or remodeling activities cause the asbestos-containing materials to crumble, flake, or deteriorate. The health effects of low exposures to asbestos are uncertain, but experts can't provide assurances that even a small level of exposure is completely safe.

According to the EPA, houses constructed in the United States during the past 20 years are less likely to contain asbestos products than those built earlier. Asbestos is sometimes found around pipes, furnaces, ductwork, and beams, and in some vinyl flooring materials, ceiling tiles, exterior roofing, and wallboard products.

CAUTION: *If you suspect asbestos has been used in your basement, attic, or garage, have the area inspected by a professional before remodeling. Never attempt to remove asbestos yourself. The materials that contain asbestos must be removed and disposed of according to strict guidelines and only by trained specialists. You'll find these experts in the Yellow Pages under "Asbestos Removal and Abatement."*

ened tip to poke wood, eliminating the need for a ladder. Infested areas must be treated by a professional exterminator before you can begin remodeling work.

Sagging Joists. Sight across the underside of the floor joists to see whether they are out of line. Those that are out of line probably are damaged but most likely can be repaired. If all the joists sag noticeably, it may be that they're improperly supported or undersized. In either case, a remedy is available.

Surveying the Garage

Surveying a garage for conversion into living space entails the same basic strategies as those for attics and basements. In addition, however, you must take into account how you'll remodel the garage door space to make it blend in with the rest of your house. You'll have to contend with a driveway that, instead of leading to a garage door, aims straight for a new wall. Because that will look odd at best, you might consider building a carport off that part of the house.

The level of the floor in the garage must also be taken into account; how much lower is it than the floor in the rest of the house? Do you want the garage conversion floor to be at the same level as the house? If so, will there be enough headroom to support a new subfloor in the garage?

You must also realize that garage floors slope about $\frac{1}{8}$ inch per foot toward the garage door opening. How will such a slope affect the usability of the conversion? The floor can be leveled out with shims and sleepers, or you can install a new subfloor that offers enough space between joists for insulation.

Garages typically have a short foundation wall that surrounds the perimeter of the space. In most cases, the wall intrudes into the garage 2 to 3 inches. Cutting out that section of the foundation is not a reasonable option. Rather, use the lip to support the framework for a new subfloor or thicken the walls with furring strips nailed to the existing studs and surface them with new drywall. The walls will come out the same distance as the foundation to make them look like the rest of the walls in your house.

Planning for Utilities

Planning the Heating System

As part of a preconstruction review of your attic, basement, or garage conversion, give some thought to how these spaces will be heated. Though it might seem tempting simply to cut vents in the ceiling below and let heat rise to an attic space or let the earth surrounding a basement act as a natural insulator, neither will work. Your home's existing heating system can usually be extended to the attic, basement, or garage if you have one of the following systems; forced-air heat (oil, gas, or electric), electric baseboard heat, or hydronic (hot water) baseboard heat.

Electric Heaters. In many cases, an electric baseboard or fan-forced electric heater is enough to supply all the necessary heat for a modest attic, separate rooms in a large basement, and most medium-sized garages. Electric baseboard and fan-forced wall-mounted heaters can be installed in home conversions regardless of the kind of heating system that already

Sagging Joists. *Sight across the underside of the joists to spot those that are out of line. Then check to see whether the whole floor system is sagging.*

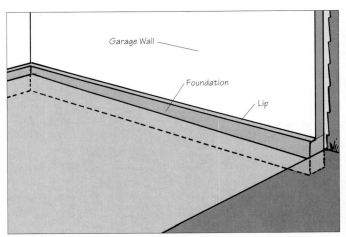

Surveying the Garage. *Use the foundation lip to guide the installation of a new wood subfloor. Note that the concrete floor slopes about $\frac{1}{8}$ in. per foot from back to front.*

exists in the rest of the house. The electric service panel, however, must be able to accommodate the additional load. Contact a heating contractor for advice before applying for a building permit. He or she will be able to suggest a suitable heater type and size and make recommendations with regard to routing heating ducts or pipes before new flooring, walls, and ceilings are installed in attics, basements, or garages.

Planning the Electrical System

Although it's possible to extend an existing electrical circuit to an attic, basement, or garage, doing so may overload the circuit. Extending a circuit that already exists doesn't provide enough power for most conversions. Plan to run at least one new circuit to an attic conversion. Although most basements and garages probably already have at least one circuit, it's best to add one or two to ensure against overloading. If you're building a home office in the attic, basement, or garage, plan to add two or more circuits and remember to have a phone line installed. Electric heaters must be served by a separate circuit. All electrical circuits in attics, basements, and garages must meet the same electrical code requirements that govern other living spaces in the house.

To accommodate the added electrical load, service to the house must be at least 100 amperes. If your house has 200-ampere service, as most newer homes do, adding the new circuits is easy. Newer homes also have three-wire service. Most homes built before 1941 have two-wire electric service, which may limit the number and type of electrical appliances that can be used. Consult a licensed electrician to determine whether the current system can be added to, modified, or upgraded.

Planning the Plumbing System

Hot-Water Supply. If your water comes from a well, the ability of the system to support a new bathroom is subject to the capacity of the pump and well. A plumbing system supplied by municipal sources can generally accommodate the addition of another bathroom. Because the water heater is probably quite a distance from the attic and may be far from the garage, you might want to add a point-of-use tankless water heater. These units are essentially miniature boilers that operate on demand. They're small enough to fit beneath the sink or in a storage cabinet. Check with a plumber for information on the availability, power requirements, and capacity of these units. Hot water should not be a problem for basements and some garages, as water heaters are commonly located in one of these two locations.

Drainage and Venting. Draining waste water and sewage is accomplished through a network of pipes that leads to the sewer or septic tank. For these pipes to drain freely, they must be connected to a system of vent pipes that leads up to and through the roof. New fixtures in an attic bathroom can usually tap into the existing system if you plan the bathroom location accordingly. Try to locate the new bathroom as close as possible to existing drainpipes, such as above an existing bathroom. The best location may depend on the direction in which the attic floor joists run. It's best to run drain lines parallel to the joists so you can avoid having to cut through, and weaken, the joists.

Basements and garages pose special problems with regard to bathrooms. Basements below grade may require special pumps that force waste and drainage up from the basement floor

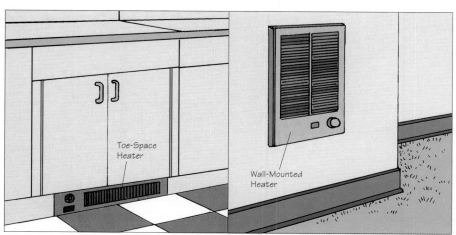

Electric Heaters. *When space is tight, you can install a heater in the toe space of a cabinet (left). A properly sized heater mounted in a wall can heat an attic space or smaller rooms partitioned off from large basement or garage conversions (right).*

Hot-Water Supply. *An electric point-of-use tankless water heater heats water only on demand.*

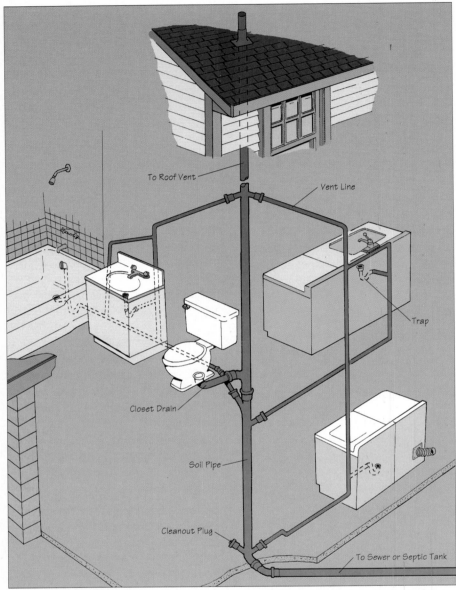

To Roof Vent

Vent Line

Trap

Closet Drain

Soil Pipe

Cleanout Plug

To Sewer or Septic Tank

Drainage and Venting. *The drain, waste, vent (DWV) system transports waste from plumbing fixtures and appliances to the sewer or septic tank.*

area to existing main drain lines leading to the sewer or septic system. For garage conversions, try to locate a new bathroom on an existing wall that already contains plumbing lines serving a bathroom on the other side. Or plan a bathroom on an exterior wall where it will be easier to run drainpipes under the foundation and to the sewer. In basements and garages, you may have to hire a professional concrete-cutting company to cut a trench in the concrete floor to accommodate drainpipes.

Who Will Do the Work?

Building codes usually allow a homeowner to work on or add to every part of his or her house, including the plumbing and electrical systems. Depending on the magnitude of the project, however, you may want to turn part or all of the work over to a professional. In addition, if you're

borrowing money for the project, the lender may require a professional to complete certain parts of the job. Be sure to question lenders on their policies.

If you want someone else to do all of the work, including locating and negotiating with each specialty contractor, such as the plumber, the electrician, the carpenter, and the carpet installer, then hire a builder or general contractor. A builder generally works with his or her own team of specialists, while a general contractor works with various independent subcontractors. You may choose to take on the role of general contractor. That means the job of hiring each subcontractor is in your hands. Working as the general contractor sometimes saves money but can be surprisingly time consuming.

Getting Bids

Screen each professional before hiring them to work on your house. Ask for references and call several people for whom he or she recently worked. The larger job, the more important it is to shop around. Get at least three bids for every phase of work. Keep in mind that the lowest bid is not always the best deal. Make sure that the professional can do the job when and how you want it, and be sure to get all agreements in writing. Provide a simple set of plans or sketches to all bidders, including each type and grade of material required, to ensure that all parties understand exactly what the job entails and can bid accordingly.

As the job moves along, you might decide to make some minor design changes. If so, talk to the contractor and come to an agreement on the cost of the change. Then put it in writing. This is called a change order, and it helps to prevent misunderstandings between contractor and homeowner.

Abiding by Building Codes

Building regulations have been around since at least 2000 B.C., when the Code of Hammurabi mandated death to the son of a builder whose building collapsed and killed the son of its owner. Codes these days are not so severe, but they do have something in common with their predecessor in that they reflect the fundamental duty of government to protect the general health, safety, and welfare of its citizens.

Attic, basement, and garage construction is covered by the same building codes that apply to work done elsewhere in the house. The codes are published in book form. You may be able to purchase the book (or codes) from your local building department. The codes can sometimes be found in the reference section of the local library.

Knowing the Codes

The codes in your community might cover everything from the way the house is used to the materials you can use for building or remodeling it. Some codes prohibit the attic of a two-story house from being used for living space unless extra steps are taken to protect it from fire hazard. Other communities may have adopted some types of fire codes, accessibility codes (requiring barrier-free access to buildings), or special construction codes, such as those requiring earthquake-resistance construction. Contact your local building department to learn the combination of codes that applies to your area.

There are several codes about which you must be aware when tackling an attic, basement, or garage con-

version. Depending on the project's complexity, some or all of the codes may affect it.

Building Code. This covers such things as the suitability of construction materials, the span of floor joists, the amount of insulation needed in a ceiling, the kind and number of fasteners used to fasten sheathing, and the amount of light and ventilation necessary to provide a healthy living space. Building codes also cover some aspects of plumbing and wiring installation.

Mechanical Code. The installation of heating and cooling equipment

(including ducts), wood stoves, and chimneys is covered here.

Plumbing Code. This code contains rules for installing water-supply and drain, waste, vent (DWV) systems, as well as related systems.

Energy Codes. In response to energy shortages in the 1970s, many municipalities instituted codes that map out minimum requirements for window glazing, insulation, and general energy efficiency.

Electrical Code. This code covers the proper installation of household electrical equipment and wiring sys-

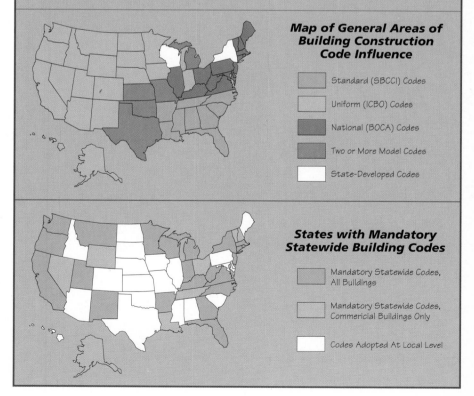

About Building Codes

Not all of the United States is covered by the same building code. In fact, each state, county, city, and town adopts those codes that best suit local building conditions. Most of the references to building codes in this book are to the 1992 edition (most recent as of this printing) of the One- and Two-Family Dwelling Code, published by The Council of American Building Officials (CABO). Though this code is widely recognized, not all towns have adopted it, and those that have might use an earlier or later version. Before you build, check with your local building officials to determine the specific codes used in your area.

Map of General Areas of Building Construction Code Influence

- Standard (SBCCI) Codes
- Uniform (ICBO) Codes
- National (BOCA) Codes
- Two or More Model Codes
- State-Developed Codes

States with Mandatory Statewide Building Codes

- Mandatory Statewide Codes, All Buildings
- Mandatory Statewide Codes, Commercial Buildings Only
- Codes Adopted At Local Level

tems. The National Electric Code, unlike most other codes, pertains to the entire country.

Following the Codes

Three regional and one national organization have developed different "model" building codes. These model codes serve as the basis for most state and local codes.

▶ The Uniform Building Code (UBC) is published by the International Conference of Building Officials (ICBO).

▶ The Standard Building Code is published by the Southern Building Code Congress International (SBCCI).

▶ The National Code, which despite the name is essentially a regional code like the others, is published by the Building Officials and Code Administrators International (BOCA).

▶ The One- and Two-Family Dwelling Code is published by the Council of American Building Officials (CABO).

Obtaining a Permit

Depending on the scope of the work, a permit application will include the following items:

A Legal Description of the Property. You can get this from city or county records or directly from your deed.

A Drawing of Proposed Changes. This drawing need not be done by an architect but must clearly show the structural changes you plan to make. It must also identify the type and dimension of all materials. Most building departments accept attic, basement, and garage conversion plans drawn by a homeowner as long as the details are clearly labeled. Note the dimensions and span of existing and new materials.

A Site Plan Drawing. This shows the position of the house on the lot, and the approximate location of adjacent houses. It also shows the location of the well and septic system, if any.

Each state decides which model codes to follow. In some cases a state may adopt two model codes and let towns pick the ones they want to use, which is why codes sometimes vary from community to community within the same state. If you have a choice, you'll probably find the CABO code easiest to understand because it's more specific about what can—and can't—be done. The other model codes, however, offer more latitude for solving unusual problems you may encounter. To determine the combination of codes that apply in your area, start by inquiring locally.

City or Town Level. Check with the local building and zoning department, if there is one, or with the housing department or town clerk.

County Level. If you live outside the boundaries of a city or town, check with the county clerk or county commission.

State Level. If you can't find a city or county office that covers building codes, check with the state offices. Codes may be administered by the departments of housing, community affairs, or building standards, or even by the labor department, which sometimes regulates builders. You might also check with the state fire marshal or state energy office.

Do You Need a Permit? Permits and inspections are a way of enforcing the building codes. Essentially, a permit is the license that gives you permission to do the work, and an inspection ensures that you did the work according to code. Usually a permit is not necessary for minor repair or remodeling work, but you may need one for adding a dormer, extending the water supply and DWV system, or adding an electrical circuit. You almost always need one to convert an attic, basement, or garage into living space.

Inspections. When a permit is required, a city or county building inspector has to examine the work. He or she checks to see that the work meets or exceeds the building codes. At the time you obtain a permit, ask about the inspection schedule. With small projects an inspector might require only a final inspection; with a larger project several intermediate inspections may be necessary before a final inspection is done. In any case, it's your responsibility to call for the inspection. It's not the inspector's job to figure out when you might be ready for a visit.

Zoning Ordinances. Another kind of regulation that can affect an attic, basement, or garage conversion is called a zoning ordinance. Some residential zoning ordinances are designed to keep multi-family homes out of single-family neighborhoods. If your attic, basement, or garage conversion plans call for the addition of a small bathroom and a separate outside entrance, local zoning officials might interpret this as an attempt to add a rental unit and may deny a permit. If you change the plans to incorporate internal stairs and access, as opposed to external, however, your conversion will no longer appear to be part of a rental unit and your plans, barring other problems, will probably be approved. Zoning ordinances sometimes restrict the height of a house or your ability to change its exterior, affecting your plans to add a dormer to the attic or to remove garage doors. If you want to add bedrooms, zoning ordinances might require you to enlarge your septic system. Another bedroom implies another resident, which in turn implies increased demand on the septic system. Though most attic, basement, and garage conversions don't run afoul of zoning ordinances, it's always a good idea to check with local officials before doing any work.

Preparation Work 3

All major remodeling projects call for a significant amount of preparation before the tasks that really make a difference can be done. You may have to remove an existing wall to make room for a new attic stairway or reinforce the attic floor to make it safe for walking. You must make sure that the structure of a basement or garage is sound and that the space is free of any moisture problems before starting a conversion. If the big problems are solved first, the rest of the project should proceed smoothly.

Planning the Logistics

Turning your attic, basement, or garage into livable space calls for a surprisingly large volume of materials and introduces some unusual logistical issues. Even small projects require flooring, lengths of baseboard, sheets of drywall, and buckets of joint compound. Getting all of these materials into the attic, basement, or garage can be tricky.

Attic

If you're opening a gable end for new windows, check with materials suppliers to see whether they use a boom truck for the main delivery. Boom trucks have a miniature crane that lifts materials directly into the attic. If your order is large enough, the company might not charge you for this service.

Catwalks. Before you spend much time working in the attic, nail some temporary flooring to the joists for your own safety. The catwalk need not be more than 24 or 36 inches wide, but it must run the length of the attic. The catwalk prevents you

Catwalks. *A platform of 1x6s keeps you from stepping through the ceiling. Be sure a joist supports the ends of each board.*

from tripping on the floor joists as you prepare the attic for construction. More than one do-it-yourselfer has ended up with bruised shins and a big hole to patch after accidentally stepping through the ceiling of the room below. Always secure the temporary flooring with enough nails to prevent it from moving as you walk. Never lay it loosely.

Lumber Transport. If your attic requires support beams, joists, and rafter stock, it probably needs them in unwieldy lengths. The most direct route to the attic isn't always the best when you're carrying 12-foot 2x8s. Look for the route that has the fewest turns. If you buy longer lengths than needed, cut them to approximate length on the ground before taking them upstairs. You may be able to slide lumber piece by piece through a gable-end window.

Subflooring and Paneling Transport. Plywood and particleboard usually come in 4x8-foot sheets. This makes them awkward to carry through the house; particularly up the stairs. Cut the panels on the ground, if possible, to fit as needed in the attic. Short kneewalls are common in attic renovation projects and offer a good chance to apply this technique. Subflooring panels must be used full size, however, to maintain their structural integrity. If you need a lot of panels, take them up in small groups over a period of several days and install each group before bringing in the next one. This doesn't save you time, but it reduces back strain and minimizes concentrated weight on the attic floor.

Drywall Transport. Drywall can be particularly awkward to carry up stairs. The panels are heavy and relatively fragile, and they come in 4x8-foot sheets that turn a stairway into a torture course. One way to ease the chore of carrying drywall is to use a lifting hook that fits around

the bottom of a sheet. If you can't find one at a local materials supplier, make one from scrap metal.

Tub-and-Shower Unit Transport. Getting a tub-and-shower unit into the attic may take some careful planning. Depending on your project, you might be able to use a crane to bring it through the rough opening in the roof just before you frame the dormer or through a gable end when you frame for a new window. Consider tiling the surround if you can't get a one-piece unit upstairs. You may also choose to use one of the tub-surround kits designed for remodeling work. These kits are made with interlocking acrylic panels that fold flat for transport.

Basement

If your basement has an exterior door, you should have few problems bringing materials into the work area. However, if yours is equipped with only an interior door, you'll face many of the same problems described for attics. The problems are compounded when interior basement stairs are L- or U-shaped.

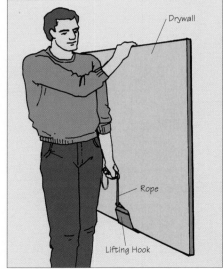

Drywall Transport. *Use a metal lifting hook to help carry sheets of drywall or plywood. The rope adjusts to suit your height.*

Consider removing drywall from both sides of the wall at the L- or U-shaped stairway landing. This way, lumber and panels can be slid through the opening and into the general basement space. Another option may be removing a basement window.

If your plans call for installing windows in a basement that has no exterior door, have the openings cut out before starting the rest of the renovation. A concrete-cutting company will have to be hired for this work. Once the window openings are made and reinforced, leave them open until all materials have been delivered through them and into the basement area. Rough openings can be temporarily secured with plywood nailed over them. Plan to frame in and install the windows after all of the large pieces of building materials, like lumber and drywall, have been brought to the basement.

Garage

Bringing building materials into a garage is easy through the big garage door. Once the opening has been framed in, however, your access will be limited. Therefore, leave one or two studs out of the door opening to make access easier. Those studs can be toe-nailed in later, after the bulk of materials has been brought inside and installed. If you're planning to install a bathtub or one-piece tub-and-shower unit in a garage conversion, be sure to bring the unit into the garage before completely enclosing the garage door opening.

Wall Removal

Nearly every remodeling project calls for the removal of existing walls or surfaces before the rest of the work can begin. Sometimes, a wall beneath an attic must be removed to make room for a new stairway. Old partition walls in a basement may need to be removed to open up the space for a new recreation room. Although garages are generally free of extra walls, previous owners may have enclosed a small section for a darkroom or hobby area. Whatever the reason, demolition is likely to be part of your project.

Putting a house together can be hazardous, but taking one apart, even partially, calls for particular vigilance. Many accidents occur simply because people expect demolition to be easy. It's not. Be as careful as you would be with any other construction project. In addition to the general safety tips listed at the front of this book, pay particular attention to the guidelines under "Demolition Safety" (left).

Removing Drywall

Electrical wire can generally be fished through walls, floors, and ceilings with a minimal need to remove sections of drywall or plaster to gain access around framing members. Running plumbing pipes, on the other hand, will require the removal of drywall or plaster because plumbing is bulkier. Many times, especially with large projects, you're best off removing all the drywall or plaster on a wall to obtain clear and unobstructed access to the wall cavities. In some cases, you might have to remove drywall or plaster only along the bottom 24 or 36 inches of a wall to gain access for electrical wiring and plumbing pipes.

Demolition Safety

■ Wear work boots. Not only do work boots help protect your feet from debris, they also shield your ankles from the scrapes and cuts commonly caused by demolition work. Wear long pants to protect your legs.

■ Wear leather work gloves. Gloves that have cuffs to protect your wrists are best. Canvas gloves can be pierced easily and are not good for this type of work.

■ Wear a dust mask and change it frequently. Even a small amount of demolition kicks up a lot of dust.

■ Always wear safety glasses—particularly when using nippers to cut nails.

■ Remove nails promptly. Taking nails out of pieces of wood or pounding them flat as you remove the wood itself prevents you from stepping on them as work progresses.

■ Frequently clean up the area and remove excess debris.

■ Proceed methodically. Remove materials piece by piece and layer by layer.

■ Never remove a wall until you know whether it's a bearing wall. Always assume that there's wiring or plumbing in the wall even if it's not evident.

■ Be careful around materials suspected of containing asbestos. If you think you've encountered asbestos, contact your health or consumer product agencies at the local or state level. Asbestos is a fibrous mineral that at one time was used as an ingredient in a variety of construction products. Now it's considered to be a serious health hazard, and experts are unable to provide assurances that any level of exposure to it is completely safe. Houses younger than about 20 years old probably don't contain asbestos. An older house may or may not contain it. Remodeling work can damage asbestos-containing materials and cause asbestos fibers to be released into the air. Places where you might find asbestos include around furnaces; in some vinyl flooring materials; in some ceiling tiles; in exterior roofing and siding shingles; in some drywall; troweled or sprayed around pipes, ductwork, and beams; in patching compounds or textured paints; and in door gaskets on stoves, furnaces, and ovens. The repair or removal of products that contain asbestos must be done by a trained contractor.

The most common indoor wall surface is ½-inch-thick drywall. The material itself isn't particularly tough, but the number of nails or screws used to attach it to the framing makes it awkward and messy to remove.

Bang a starter hole through the old drywall with the claw of a hammer or with a pry bar tapped with the hammer. Use the claw of the hammer or pry bar to pull off a large chunk at a time, removing all nails as you go. Protect your hands with gloves when removing metal trim edges.

Removing Plaster and Lath

Stripping plaster from a wall is dirty work no matter how it's done. Some people prefer to cut it away in chunks using a circular saw with a masonry blade set to make a shallow cut. Others find it easier just to batter the plaster with a hammer, pull it away in chunks and pry away the lath using a wrecking bar. In either case, be certain to wear safety goggles and a quality dust mask.

Removing Framing

There are two basic types of walls in every house, and it's essential that you identify each one before attempting to remove it. If you skip this step, you risk injury to yourself and serious damage to your house.

With regard to attic and basement conversions, it's most likely that you may want to remove a wall upstairs to make way for a new attic stairway or downstairs to open up the entire basement space. Most garages are wide open, and you shouldn't need to remove any walls. In attics and basements, make certain that any wall you want to remove isn't a bearing wall.

Bearing Walls. A wall that supports structural loads, such as a floor, a roof, or another wall above, and

Removing Drywall. *Tear out drywall by using the claw end of a hammer or pry-bar; remove all nails as you go.*

Removing Plaster and Lath. *Pulling plaster from a wall is messy business. Wear goggles, a dust mask, and gloves.*

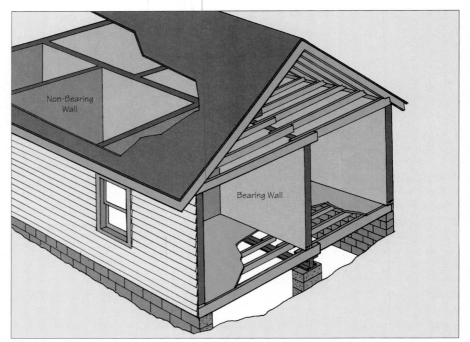

Bearing Walls. *Before removing a wall, determine whether it's a bearing wall. Look for clues such as lapped joists above. Walls that are parallel to the attic joists usually are not load bearing.*

helps to transmit those loads to the foundation of the house is a bearing wall. Except for gable end walls, most exterior walls are bearing walls. Usually a wall that runs lengthwise through the center of a house is a bearing wall. Joists that run along each side of the house rest on the center bearing wall. Bearing walls sometimes can be spotted from the attic. Look for two sets of overlapping joists. The wall on which the ends rest is a bearing wall. You also

may be able to identify bearing walls from the basement. Look for walls that rest atop a beam or a basement wall. If you're not sure about the kind of wall you're dealing with, the safest thing is to assume that it's a bearing wall. To remove bearing walls safely, you must identify all the loads involved and provide alternative support for them. Seek professional advice from a builder or engineer when you must remove a bearing wall.

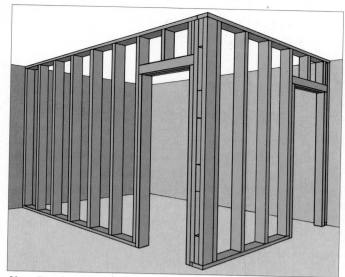

Non-Bearing Walls. *New walls built inside an existing structure like a basement or garage won't be load bearing; they'll be partition walls designed simply to separate certain areas from others.*

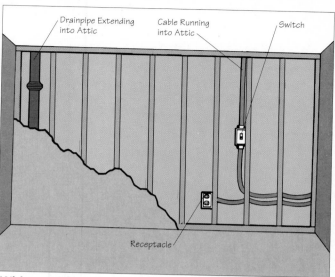

Drainpipe Extending into Attic

Cable Running into Attic

Switch

Receptacle

Wiring and Plumbing. *Wiring and plumbing are two commonly encountered utilities within a wall. Examine the wall from the attic and the basement to determine the utilities' exact locations.*

Non-Bearing Walls. Non-bearing walls, also called partition walls, support only the wall covering attached to them. Usually a non-bearing wall can be removed without affecting the structural integrity of the house. If a wall doesn't support joist ends and doesn't lie directly beneath a post, it may be a non-bearing wall. Walls that run parallel to ceiling joists are usually non-bearing walls.

Wiring and Plumbing. Before removing a wall, check the area immediately above and beneath it from the attic and basement, if possible. Look for wires, pipes, or ducts that lead into the wall. There's no way to tell for sure how big the job is until you pull the drywall or plaster from at least one side of the wall.

Wiring is easy to relocate, but water supply piping (hot- and cold-water pipes) is more difficult. Plumbing vent pipes are trickier still, primarily because of code requirements that restrict their placement. Heating ducts and drainpipes are the toughest of all to relocate. Consult a professional if you're unsure of what to do. With luck, the wall you open up will contain nothing but dust.

1 *Use a sledge hammer to loosen one stud at a time.*

Removing a Non-Bearing Wall

Difficulty Level: 🔨🔨

Tools and Materials

☐ Basic carpentry tools
☐ Sledge hammer
☐ Wrecking bar

1 **Loosen the Studs.** After stripping the drywall or plaster from both sides of the wall and relocat-

2 *Twist each stud away from the top wall plate.*

ing utilities, use a sledge hammer to force the bottom of each stud away from the nails that hold it in place. Proceed cautiously until you get the feel for how much force you need to exert.

2 **Remove the Studs.** Once the bottom is loose, grasp the stud and push it first sideways, then outward and back and forth to work the top nails loose until you can remove the stud. Another way is to cut and remove studs in smaller sec-

3 *Use a wrecking bar to lever up the bottom plate.*

tions. Either way, loosen or cut and remove one stud at a time or you'll find yourself assaulted by a dangling row of loose or severed 2x4 studs. As you work, flatten nails that protrude from the bottom plate so they can't puncture your foot as you work on the next stud.

3 **Pry Off the Plates.** Use a wrecking bar to remove the end studs, then the top and bottom plates. Nails may be located anywhere along the length of each end stud.

Removing Nails

There'll be a lot of nails to remove, and when they're removed correctly a potential hazard will be eliminated. You can save money by reusing some old materials such as trim and molding. Nails can't be reused.

Finishing Nails. It's best to remove the molding itself first, then the nails. To remove the molding without damaging it, place a scrap of wood behind the pry bar (to protect the wall surface) and gradually pry the molding away from the wall. If the nails poke through the front of the molding, use a pry bar to remove them. If the nails stay in the molding, as they often do, the best way to remove them is with nippers. Grasp the shank of the nail, and lever it from the back side of

the molding. In a pinch you can do the same thing with side cutters. The beauty of this technique is that there are no nailholes to patch: The nail pulls through the back of the wood without disturbing the face.

Large Nails. Nails are not always easy to remove. Sometimes the best you can do is to cut a nail flush with the surface of the wood. This is another job for nippers. Grasp the nail as if you were going to pull it out, then squeeze the nippers as hard as you can. Small nails shear off easily. However, the shank of larger nails (about 10d and above) may not yield to one bite of the nippers. If this is the case, grip the nail hard, then rotate the nippers one-quarter turn and grip again. This creates a score line on the nail. Move the nippers higher on the exposed nail shank and bend the nail over, shearing it at the score line.

Use a pry bar to pull nails from framing lumber. Place the notch at a slight angle to the surface of the wood, then strike the pry bar just hard enough to drive the notch under the head of the nail. You may need to hit it a second time to ensure a good grip on the nail shank. Use one smooth motion to lever out the nail. Use a cat's paw in a similar fashion.

Large Nails. *Use a pry bar or cat's paw and hammer to remove nails that have sunk below the surface of the wood.*

Removing Collar Ties and Kneewalls

Always consider the collar ties and kneewalls you find in an unfinished attic to be structural elements, just to be on the safe side. If the success of an attic renovation absolutely depends on moving collar ties or structural kneewalls, consult an engineer or architect for advice.

Collar Ties. The weight of the roofing, along with wind and snow loads, pushes down on the rafters. Collar ties hold the ridgeboard and rafters together, and the attic floor joists keep the exterior walls from being pushed outward. The position, dimension, and number of collar ties determine their effectiveness.

Kneewalls. Sometimes kneewalls are used to support rafters at mid-span. If a house is particularly wide, for example, the rafters would have to be unusually hefty to reach from the ridge to the outer walls. With a kneewall, however, the span of rafters can, in effect, be cut in half and a smaller dimension of lumber be used. The kneewall usually transfers the load to a wall or beam beneath. By moving the kneewall, you change the effective span of the rafters and might transfer loads to a part of the house that can't handle them.

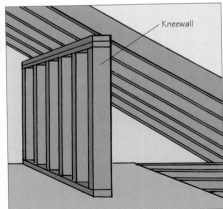

Kneewalls. *These short walls may be non-structural, or they may support rafters at the middle of their span.*

Garage Door Removal

The ultimate goal of every home remodeling project is to make the remodel appear as though it were part of the original house. Attic and basement conversions pose little problem in this regard compared with garage conversions. Garages have a driveway that runs up to one or more large, conspicuous doors. If you remove the garage doors as part of the conversion, the driveway makes the house look awkward. This problem can be solved by erecting a carport in front of the garage conversion to give the driveway some meaning. Otherwise, you should seriously consider removing all of the driveway and then landscaping that area to match the rest of your yard's landscape. In some cases you might be able to remove a 6- or 8-foot section of the driveway next to the house and install planters or a flower bed. The goal is to make the house appear as if it never had a garage in the first place.

Removing an Electric Garage Door Opener

Unplug the opener's electrical cord from the receptacle and disconnect the wires that lead to the fixed operating switch. Pull the cord that releases the door carriage arm from the track to relieve any tension on it. Then disconnect the carriage arm from the garage door. With a helper, loosen the bolts that secure the carrier track to the motor housing, then carefully loosen the bolts that secure the other end of the track to the wall or ceiling adjacent to the garage door. Maneuver the carrier until the chain is loose enough to come off the sprocket at the motor. The entire carrier assembly should

come down in one piece, which is why you'll need a helper. Loosen the bolts that secure the motor to the ceiling, and again with a helper, lower the motor to the floor. Loosen and remove all of the other bolts that secure brackets and other hardware to the ceiling and the garage door.

Removing Swing-Up Overhead Doors

Swing-up overhead doors are one-piece units that pull out from the bottom and swing up to rest horizontally above the inside of the garage door opening. They're heavy

Garage Door Removal. *A driveway that goes to a garage looks normal. After a garage conversion has been built, however, the driveway may appear awkward as it leads straight to a wall (instead of a garage door). To justify the driveway, you can build a carport in front of the new garage conversion.*

Carriage Arm

Carrier Track

Opener Motor

Removing an Electric Garage Door Opener. *Unplug the unit from its power supply. Pull the release cord and disconnect the carriage arm from the garage door. With the arm free, loosen the bolts that secure it to the door and remove it completely. With a helper, loosen and remove the bolts that secure the carrier track to the ceiling or garage-door wall. The track is long and heavy; once it's loose, maneuver it so you can free the chain that fits over the motor sprocket.*

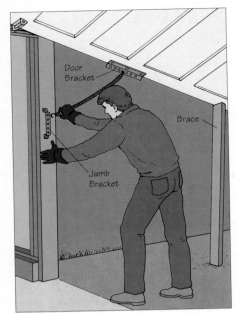

Removing Swing-Up Overhead Doors.
Springs assist in lifting open a single-piece swing-up garage door. When the door is open, spring tension is released and the springs are easy to remove. The door must be braced in the open position with 2x4 boards before you can dismantle the springs.

and require sturdy springs on each side to help get them raised and hold them in an open overhead position. Note that the springs are extended when the door is down and compressed when the door is up. Minimal spring pressure is applied when doors are in the up position, and maximum spring pressure is exerted when doors are closed.

Open the door and wedge 2x4s between the door bottom and the floor. Be certain the door is securely braced before continuing. With the door wedged open in this position, much of the tension should be off the springs. Disconnect the springs from their fittings and place them on the floor away from your work area. With at least two helpers (more for extra-wide garage doors) standing at the front of the door and holding up the bottom edge, carefully remove the 2x4 braces. The door will be heavy and may try to fall to the

closed position by itself. With all braces removed, carefully lower the door to the ground and remove all the hardware around the door jamb and frame.

Removing Roll-Up Sectional Overhead Doors

Sectional overhead garage doors are probably the most common. The doors consist of a number of horizontal panels secured with hinges that have rollers protruding from their sides. The rollers follow a curved track that leads them to a horizontal position above and inside the garage door opening. Sectional doors use two kinds of lifting systems, extension springs and torsion springs. Extension springs flank the door and are attached through a cable-and-pulley system. The springs are extended—under tension—when the door is closed and take most of the weight of the door when you open it. With a torsion-spring system, a large coiled spring is mounted horizontally across the top of the opening and is used to assist in opening the door through cables that stretch from the spring and attach to the bottom of the door.

Extension Spring Removal. To remove extension springs, open the

door and keep it open with locking pliers clamped to the roller track. With the door open, the springs are relaxed enough for you to remove the lift-cable pulley. Once you've removed the pulley, unhook the extension springs and detach all the cable connection points.

Torsion Spring Removal.
Removing a torsion spring requires careful unwinding of the coiled spring. This process is difficult and extremely dangerous, as the spring is under a tremendous amount of tension when the door is closed—the only time the spring is accessible. If you're not a skilled mechanic, have a professional garage-door service technician remove the spring. If you decide to do it yourself, you'll need a heavy bar, a wrench, and a helper to slowly and cautiously unwind the spring. With the garage door closed, insert the bar into one of the holes in the round bracket at the outer end of the spring. Hold the bar firmly and put tension on the spring as your helper slightly loosens the locking bolt. Then, unwind the spring one-half turn or as much as the bar will allow. Have your helper tighten the locking bolt, then reposition the bar into a different hole in the round spring bracket. There are about four holes in the bracket to accommodate this process. Follow the process

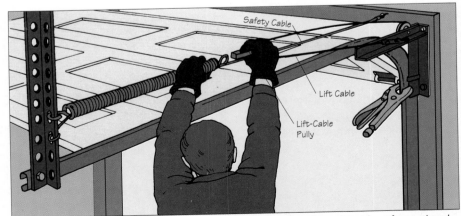

Extension Spring Removal. *To remove extension springs from the sides of a sectional overhead garage door, hold the door fully open with locking pliers, disconnect the lift cables, and unhook the pulley from the spring.*

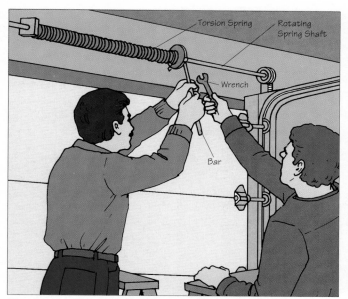

Torsion Spring Removal. *A single torsion spring may be mounted horizontally on the header. The spring must be unwound with a bar when the door is closed. This is dangerous and should be done by a professional.*

Door Removal. *Once you've removed the springs, cables, and pulleys, detatch the hardware from the top section and remove it. Remove the succeeding sections, then the tracks and other hardware.*

slowly and carefully, one-half turn at a time, until you release all of the tension from the spring.

Door Removal. Once all of the tension has been released from the torsion or extension springs, remove them along with the cables and pulleys. Unbolt the hinges and roller brackets from the top horizontal door section and remove it. With that section removed, continue until you've removed all the sections from the tracks. Then unbolt the tracks and other hardware.

Joists and Rafters

It may be likely that a joist or rafter in an older home became damaged over the years. Maybe a small crack turned into a large one, or perhaps someone cut into a joist or rafter for reasons that made sense at the time. Damaged joists or rafters must be repaired before you install a new ceiling (new floor joists above for basements and rafters for attics and garages). Not all cracks compromise

the strength of a joist or rafter, but if one is sagging or if a crack runs clear to the bottom edge of the board, repair is in order. Only in rare cases should you remove and replace a joist or rafter, even if one is seriously damaged. Remember, the floor sheathing is nailed to the floor joists and roof sheathing is nailed to the rafters, so pulling

either risks damaging the finished floor or roof above.

Repairing a Joist or Rafter

You can reinforce or repair an existing joist or rafter by attaching an equal-sized piece of lumber alongside of it. This process is called

Repairing a Joist or Rafter. *The sister board must be the same dimensions as the joist it supports; cut off one long corner to maneuver it into position.*

1 The bottom of the ceiling joist is installed ¹/₂ in. above the finished ceiling. Make sure the ceiling joists are level.

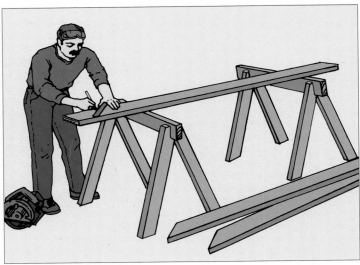

2 Joists must be angled on each end to match the roof pitch. The cuts need not fit exactly, so don't spend too much time on them.

"sistering." The new lumber must be as long as the existing board and of the same depth, and it will be supported in the same locations as the damaged joist or rafter. Maneuver the new board into position (cutting off a portion of one corner may help), push the old joist or rafter back where it belongs, and use 16d nails to nail the two together. You'll have to shim under any cut corners for proper support.

Installing Attic Ceiling Joists

Some people prefer the look of a vaulted ceiling in an attic conversion, but a flat ceiling gives the attic a look that's more in keeping with other rooms in most homes. If you want a flat ceiling, you'll have to install ceiling joists. Basements already have ceiling joists by way of the floor joists that support the floor above. Some garages have only a few ceiling joists installed simply to hold the roof together; you may have to add a full complement of ceiling joists to install a regular ceiling. Flat ceilings allow the easy installation of lights and provide space for gable vents in attics. Use 2x6s for ceiling joists and space them 16 inches on center. If

the joists will span more than 10 feet, use 2x8s or 2x10s (determined by consulting a spanning chart at the lumberyard). If the garage attic space is large, consider using 2x10 lumber for the joists and covering it with plywood for extra storage.

Difficulty Level: 🔨🔨

Tools and Materials

☐ Basic carpentry tools
☐ Joist stock (2x6s, 2x8s, or 2x10s)
☐ Sliding T-bevel
☐ Circular saw
☐ Chalkline
☐ Common nails, 16d

1 **Locate the Joists.** Decide exactly how high you want the finished ceiling in the attic or garage to be, then add ¹/₂ or ⁵/₈ inch (depending on the thickness of the drywall you use) to determine the height of the joists. At this point, mark a horizontal layout line on a pair of opposing rafters. Measure across the opposing rafters at the level of the layout lines to determine the joist length.

2 **Cut a Template.** Use a sliding T-bevel to copy the angle of the roof onto the joist stock; then use a circular saw to cut it. Test-fit a joist;

3 Snap a level chalkline across the rafters to position the joists. Sight down the row of rafters to be sure they're in the same plane.

if it's correct, use it as a layout template for cutting the rest of the joists. Note that the fit against the underside of the roof surface need not be exact.

3 **Mark the Rafters.** Mark the height of the joists onto several additional rafters and snap a chalkline between the marks. Align the bottom of the joists to the chalk line and use two 16d nails on each end to nail the joists into the sides of the rafters. Double-check each joist for level as it's installed.

Concealing Heating Systems

A basement or garage is generally packed with remodeling obstacles that include a sump pump, a water pump, a water heater, pipes, ductwork, drains, and a heating unit, among other things. It's easiest to avoid these items from the start, so try to work your design around them. If you can't avoid the mechanicals, however, often they can be concealed.

If the central heating unit shares space with basement or garage living areas, safety is the most important issue to consider. The furnace or boiler is best located in a separate room, though it can share the space with other mechanical equipment like the water softener and water heater. Local building codes regulate the distance between heating appliances and combustible walls, so check local codes before building a room for the heating plant and other combustion appliances. Building codes may also regulate the size and details of the room; check them for specifics.

Generally, a furnace or boiler room must have a door that's big enough to remove the largest piece of equipment, though in no case can the door be less than 20 inches wide. The room must contain an unobstructed working space on the control side of the heating unit that's at least 30 inches wide and 30 inches high. It's also a good idea to provide a light controlled by a switch near the door.

Be sure to vent the room so that there's enough incoming air for a combustion-type furnace or boiler. The amount of air required depends on the type and capacity of the furnace or boiler and/or water heater; check with a local heating contractor.

Service Corridors. Oil-fired combustion appliances are supplied with fuel through copper fuel lines. These lines have a fuel filter that must be accessible for periodic maintenance. Fuel lines must never run beneath flooring but can be hidden behind a partition wall. By leaving about 24 inches between the partition wall and the foundation, you'll create a service corridor that can be used for ready access to the line. A service corridor also can be used to maintain access to other devices, such as sump pumps, water pumps, and the like.

Carbon Monoxide Detector. As a safety precaution (particularly if someone is to sleep in the area even occasionally) install a carbon monoxide detector. Carbon monoxide is a colorless, odorless, and potentially lethal gas that's a by-product of combustion. Under normal circumstances the gas is vented out of the house, but a faulty furnace, boiler, or gas- or oil-fired water heater may cause carbon monoxide to leak into the basement or garage. A detector sounds an alarm when it senses carbon monoxide.

Moisture Problems

Eliminating Attic Moisture Problems

Moisture problems in attics are generally related to two things: a leaky roof or moisture condensation owing to inadequate ventilation. Attic spaces at the underside of the roof level must have plenty of airflow to eliminate the development of condensation from warm moist air rising from the house interior. Cathedral ceilings are no exception; they must have at least a 2-inch ventilated air space between insulation and the underside of the roof. Soffit vents on at least every other opening between rafters, gable end vents, and a ridge vent should provide plenty of attic ventilation.

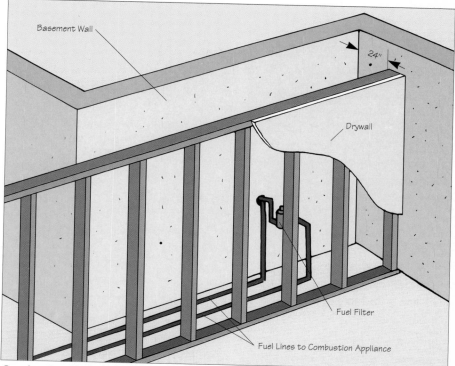

Service Corridors. *By leaving about 24 in. of space between a foundation wall and a partition, you create a service corridor that allows for easy access to various devices.*

Eliminating Garage Moisture Problems

Moisture problems in a garage may be a result of a leaky roof, water seeping in under garage doors, water-pipe condensation, or water seeping in through foundation walls. Leaky roofs must be patched or have flashing repaired. A new foundation across the garage door opening will prevent water from seeping in, and a drain just outside that opening might alleviate the problem altogether. You can solve water-pipe condensation with pipe insulation. Water seeping through foundation walls, however, is a different problem, and one that's shared with basements.

Eliminating Basement Moisture Problems

A basement, or a garage with large foundation walls, can't be turned into a suitable living space unless it's guaranteed to stay dry. Water problems range in seriousness from mild condensation and seepage to periodic flooding. Given enough time and money, all water problems can be solved. But that doesn't mean the effort is justified. Assuming that the plumbing doesn't leak, basement moisture comes from either seepage (water from outside the house leaking through walls or floor) or condensation (the result of warm moist air hitting a cold masonry wall or cold water pipes). The source of the water can be identified by performing a simple test (see "Moisture Problems," page 38).

Condensation. If condensation is the problem, eliminate it either by installing a portable dehumidifier in the basement or by insulating the walls and water pipes.

Seepage. Seepage water is more difficult to eliminate because it might be coming from any or all of the following sources:

▶ *Gutter Systems.* Leaders that dump water near the foundation encourage water to soak in at exactly the wrong places. Use splash blocks or leader extensions to direct water away from the house. Clogged gutters allow water to spill over and run down the siding toward the foundation wall.

▶ *Improper Grading.* If the grading slopes toward the house or allows water to pool near the foundation, it's a problem. To conduct water away from the foundation, the grade must drop at least 2½ to 6 inches in 10 feet all around the house. Fill in pockets that encourage water to pool.

▶ *Lack of Footing Drains.* Some water inevitably reaches the bottom of the foundation, but it will not be a problem if perforated pipes, called footing drains, lead it away. Most newer houses have footing drains, but older homes may not. Drains can be added to older houses, though not without considerable effort. This is a job for a contractor. Heavy equipment is required to excavate the entire perimeter of the foundation. The job also involves placing gravel sub-drains, placing perforated pipes for the main drains, and lastly, back-filling and regrading.

▶ *Cracked Foundation Walls.* Water finds a way to get through even the smallest cracks, so use hydraulic cement to patch all of them.

▶ *Pipes or Electrical Lines.* The problem isn't the pipe or the line, it's the gap around the pipe and the line that leads water into the basement. Seal gaps with hydraulic cement or silicone sealant (a high-performance product similar to silicone caulk).

▶ *Poorly Waterproofed Foundation.* Concrete and concrete block are not waterproof. A foundation that's not properly waterproofed allows moisture to migrate directly through the masonry. If the foundation wasn't waterproofed or if the waterproofing has failed, installing a proper waterproofing system isn't something the typical homeowner should attempt. As with installing new footing drains, a good bit of excavation is required.

▶ *Nearby Vegetation.* Plants hold moisture in the soil, and their shade reduces the evaporation of ground moisture. Both factors add to water woes. Another factor is the presence of plants that require a lot of deep watering, adding moisture to the soil that could seep into the basement.

▶ *High Water Table.* The water table varies in depth from area to area and even from season to season. Nothing can be done about the level, but foundation drains and sump pumps help conduct water away before it becomes a problem.

Sealing a Masonry Wall

Even if the basement isn't plagued by the kind of water problems that show up as active drips, moisture still may be seeping through the masonry itself. This kind of moisture movement can be stopped by sealing the walls from inside the basement. Even if the walls seem to be dry, sealing them is a reasonable precaution to take. After all, it doesn't take much moisture to warp wood paneling or to encourage a musty smell. To seal the walls, brush them with a product that contains portland cement and synthetic rubber. Suitable products go by many names: cement paint, waterproofing paint, basement paint, or basement water-proofer. Though some brands claim to keep out water that's under a modest amount of pressure, nothing applied to the inside of the walls can solve serious water problems. After applying a waterproofing paint, cover it with a quality latex paint.

Difficulty Level:

Tools and Materials

☐ Goggles
☐ Rubber gloves
☐ Wire brush
☐ Water
☐ Acid etching compound
☐ Stiff bristle brush
☐ Cold chisels (various sizes)
☐ Shop vacuum
☐ Hydraulic cement
☐ Trowel
☐ Wide nylon brush
☐ Waterproofing paint

1 Clean the Surfaces. Use a wire brush to remove loose mortar and dirt from the walls. Sealing is most effective on a wall that's never been painted, but if all of the old paint is removed, the sealant still has a chance to do its job.

2 Remove Efflorescence. A harmless, white crystalline deposit called efflorescence sometimes forms on concrete or concrete-block walls. Efflorescence is caused by water-soluble salts within the masonry that migrate to the wall's surface and interfere with the bond between the waterproofing paint and the wall. A commercial etching compound that's dissolved in water and applied with a stiff bristle brush can be used to remove efflorescence. Etching compounds contain a mild acid, however, so follow manufacturer's application and safety instructions to the letter: Rubber gloves and eye protection are mandatory. Use clean water to rinse the wall surface thoroughly so the acid is neutralized. Then let the wall dry thoroughly.

3 Prepare Cracks and Holes. Use a chisel to undercut cracks and holes slightly to provide a "key" that holds hydraulic cement in place. Hydraulic cement is particularly effective at sealing cracks where moisture is present, and because it expands slightly as it cures, it locks tightly

1 Use a wire brush to remove loose mortar and dirt from the concrete-block wall, then vacuum the wall to remove the remaining dust and debris.

2 Use a bristle brush and a mixture of etching compound and water to remove efflorescence from masonry. Wear rubber gloves and eye protection.

3 Chisel out and undercut cracks and small holes to provide a firm anchor for hydraulic cement. Clean out loose debris.

to a properly prepared crack. After keying, vacuum the crack area to remove loose dust and debris.

4 Apply Cement. Mix a small amount of powdered hydraulic cement with water. If the leak is active, wait until the mixture becomes warm to the touch (indicating that it's beginning to set), then use your hand to force it into a portion of the crack. Hold the mixture in place for several minutes until it cures. If the leak isn't

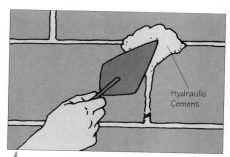

4 Mix powdered hydraulic cement with water. Use a trowel or your hand (with gloves on) to apply the cement to the damaged area.

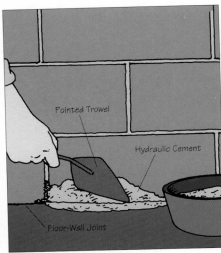

5 *Use a pointed trowel to apply hydraulic cement to the floor-wall joint.*

6 *Apply masonry waterproofer to the walls with a wide nylon paintbrush. Work the liquid into the rough surface of concrete block.*

active, soak the area with water first, then use a trowel to force the cement mixture into cracks and holes. Use the trowel to smooth out patched areas immediately.

5 Plug Other Water Problems. Water seepage may also be a problem at the juncture of walls and floor slabs. Use a liberal amount of hydraulic cement to seal the area, then smooth it with a trowel.

6 Apply Waterproofing Paint. Once the hydraulic cement has cured and the wall is dry, use waterproofing paint to seal the walls. Make sure the work area is well ventilated. For best results, use a wide nylon bristle brush to dab the material into the pores of the masonry. Allow the first coat to dry overnight, then apply a second coat.

Correcting Severe Water Problems

If water continues to enter the basement despite efforts to seal the walls from the inside, the problem must be tackled from the outside. If excessive quantities of water build up in the soil just outside the foundation, they will be forced (by hydrostatic pressure) through the masonry. Water that's under a modest amount

of pressure can be sealed out with waterproofing paint; however, large amounts of pressure can defeat any product that's applied to the inside of a wall. A waterproofing layer

applied to the outside of the foundation is far more effective.

Exterior Waterproofing. The more pressure applied to an exterior water-

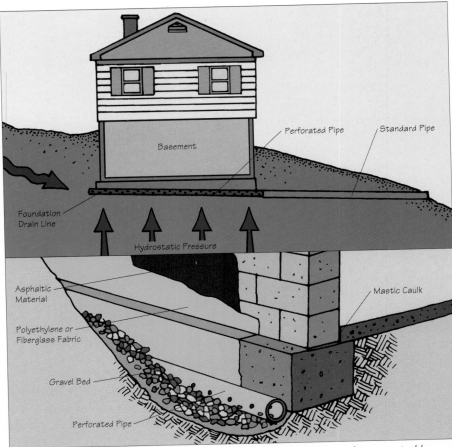

Correcting Severe Water Problems. *Severe water problems can be corrected by intercepting water before it reaches the foundation with foundation drains.*

proofing coating, the tighter the waterproofing adheres to the wall. It's not easy to waterproof the outside of the foundation, however, and it can be quite expensive. Because all possible strategies involve a good bit of excavation, this work is best left to a contractor. The work typically involves digging down to the footing, installing a system of drainpipes to redirect water around the house, and applying a protective waterproofing membrane to the outside of the foundation walls. Compare several estimates before signing a contract.

Sump Pumps. One fairly easy way to keep the basement dry is to install an electric sump pump, which draws water from beneath the slab and pumps it away from the house. The pump sits in a hole, or sump, that extends below the slab. When water collects in the sump, the pump turns on automatically and removes the water through a plastic discharge pipe that exits the basement above grade. Installing a sump pump calls for wiring, plumbing, and concrete-demolition skills. A pump must not be the only thing standing between you and a flooded basement, however. After all, pumps do fail. A sump pump may be part of a strategy to keep water out, but if water problems are severe, footing drains may have to be installed as well.

Types of Sump Pumps. There are two basic types of sump pumps:

pedestal and submersible. A pedestal-type pump features a raised motor that does not come in contact with water. Instead, it sits on top of a plastic pipe that extends into the sump. Water rising in the sump causes the float to rise and turn on the pump. When the water level drops, so does the float, which then turns off the pump. With a submersible pump, the entire pump sits at the bottom of the sump pit and is submerged every time the sump fills up with water. A float on the pump triggers the on-off switch. Either kind of sump pump removes water effectively. Consult a plumber or supplier to determine the best sump pump for your situation.

Installing a Sump Pump

In addition to plumbing codes, the pump installation must conform to electrical codes, which in general terms means that the pump's power supply must be a dedicated 15-amp circuit. The following step-by-step project is a guide to how the work usually is done, but be sure to check local codes and the manufacturer's instructions that come with the pump. The exact dimensions of the sump pit, for example, depend on the size of the liner, sometimes called a basin. Purchase the pump and the liner before starting work.

Difficulty Level: 🔨🔨 ***to*** 🔨🔨🔨

Tools and Materials

- ☐ Goggles
- ☐ Gloves
- ☐ Sump pump
- ☐ Sump pump liner
- ☐ Pencil
- ☐ Measuring tape
- ☐ Jackhammer (rented)
- ☐ Shovel
- ☐ Gravel
- ☐ PVC pipe
- ☐ Check valve
- ☐ Masonry anchors
- ☐ Pipe brackets
- ☐ Electric drill
- ☐ Masonry bits
- ☐ Hole saw for PVC pipe
- ☐ Surface-mounted outlet
- ☐ Wiring
- ☐ Basic electrical tools
- ☐ Short length of 2x4

1 Dig the Sump. The sump pump is located in the lowest part of the basement so that water naturally drains to it. The whole point of installing the device is to keep water out of the basement to begin with, of course, but it's still prudent to locate the pump this way. Turn the basin upside down and use it as a template to scribe a circle on the

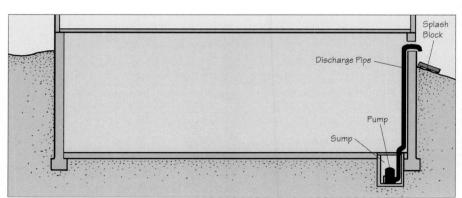

Types of Sump Pumps. *The pump motor on a submersible pump (shown) is immersed in water. The pump motor on a pedestal pump stays clear of the water.*

1 Rent a jackhammer, as shown, or a rotary hammer to reduce work.

floor; the circle's centerpoint must be about 16 inches away from both walls.

Breaking a hole in a 4-inch-thick slab isn't easy work, so rent a jack-hammer for the job. Wearing safety glasses, remove concrete up to the layout line. Then dig out the soil and gravel beneath the slab—the depth depends on the size of the liner, usually about 24 inches. Periodically, slip the liner into place to see whether it fits.

2 Install the Liner. The liner is a plastic tub with holes in the sides that allow ground water to seep into it. Some liners have a lip at the top that covers the edges of the concrete. Once the hole is deep enough, slip the liner into place and fill it with gravel, as needed. If necessary, use concrete to seal the edges of the hole.

3 Hook Up the Discharge Pipe. Check the manufacturer's instructions that came with the pump to determine the type of discharge pipe needed and the necessary diameter (usually 1¼-inch PVC pipe). Attach a piece of pipe that's long enough to reach the pump itself (submersible type) or the intake housing (pedestal type). Then lower the pump into the sump and connect additional discharge pipe as needed. A check valve must be used somewhere in the discharge piping to keep water from draining back into the sump. If the pump has to be removed for servicing, disconnect the discharge pipe at the check valve. It usually has a threaded fitting.

4 Install the Discharge Pipe. It's usually easiest to route out the discharge pipe through the rim or header joist. Cut an appropriate size hole with a drill and hole saw. Attach the pipe to the basement wall with masonry anchors and brackets. Afterward, use caulk to seal the hole in the joist. Be sure

2 Slip the sump liner into place and backfill it with gravel as needed.

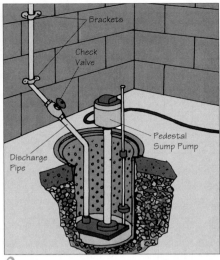

3 The discharge pipe usually is PVC and must be secured to the foundation walls with support clamps.

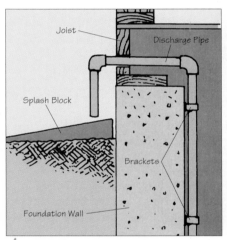

4 Cut a hole in the rim joist and route the discharge pipe through it. Direct water away from the foundation wall.

5 Use a 2x4 secured to the foundation walls with masonry anchors to provide a base for installing the electrical box.

that the end of the discharge pipe doesn't dump the water against the outside of the foundation. Use a splash block to direct water away.

5 Wire and Test the Pump. Check local electrical codes to see whether a ground fault circuit interrupter (GFCI) is required for a sump pump, then run wiring from the service panel to the general location of the new pump. The pump itself has a long electrical cord and three-prong plug, so you must install a grounded receptacle for it. This arrangement makes it easy to dis-

connect the pump should it ever need servicing.

The box is typically located high on a wall to avoid splashes caused by the pump. Check local codes for specifics. Use masonry anchors to secure a 2x4 to the foundation wall; this will provide a secure base for the installation of a surface-mounted electrical box. Plug in the pump and pour water into the sump. The pump should begin to work when the sump is about half full. As water flows through the discharge pipe, check all connections for leaks.

Building Stairs and Framing Floors

Stairways are necessary to gain access to an attic, a basement, and sometimes to a garage. Stairs aren't particularly difficult to build, but building codes are strict when it comes to their construction, in part because small variations in details like step height make a big difference in the safety of the stairs. Floor construction, for the most part, is straightforward—except for attic spaces. There you have to understand concepts like loading forces and how to deal with them to make the floor safe.

Basement Stairs

One advantage to adding new living space in a basement, rather than an attic, is that a stairway is already in place. It may be suitable just the way it is, but in some cases it may have to be rebuilt. In converting a basement, for example, you may decide to insulate the concrete floor with a system of wood sleepers and rigid insulation capped with plywood and carpeting.

The thickness of this assembly will change the height of the last step on the stairs, making it shorter than the others by the thickness of the new floor system. Unless this problem is corrected, the stairs will not be safe and the project will not pass code inspection. In this situation, the stair carriage can't simply be raised—the stairs have to be rebuilt.

There are other reasons to rebuild stairs. Although basement stairs in all newer houses have to adhere to the same codes as those anywhere else in the house, this was not always the case. If your house is an old one and the basement stairs are uncomfortably steep or poorly constructed, they must be rebuilt. Rebuilding can usually be done without enlarging the stairwell itself.

Adding a Balustrade

Many existing basement stairs may not be up to code. In particular, handrails and railings often are missing or inadequate. If the stairs are usable otherwise, however, balustrades can be easily added. Bolt the balusters directly to a stringer, then countersink the bolts or screws and conceal them with wood plugs. Space the balusters so that the opening between them is no more than 6 inches measured horizontally. The top of the handrail must be easy to grasp.

Basic Stair Dimensions

Your local building codes are the last word on stair dimensions, but the following can be used as a guide:

■ The width of the stairs must be at least 36 inches. Measure between finished walls.

■ Nosing (if used) must not project more than 1½ inches.

■ Headroom must measure at least 80 inches from the tip of the nosing to the nearest obstruction at all points on the stair.

■ The ratio of riser height to tread depth should total 18 inches. The ideal riser height, for example, is 7 inches and the ideal tread depth is 11 inches (7 + 11 = 18). For safety reasons, risers must be no more than 8¼ inches high and treads must be at least 9 inches deep.

■ All stairs made up of three or more risers must have a 30- to 38-inch-high handrail on at least one side. Handrails are measured vertically from the tip of the tread nosing. The end of the handrail must return to the wall or terminate in a newel post to mark the end of the stairs even in the dark.

■ Landings must be the same width as the stairs and at least as long as they are wide.

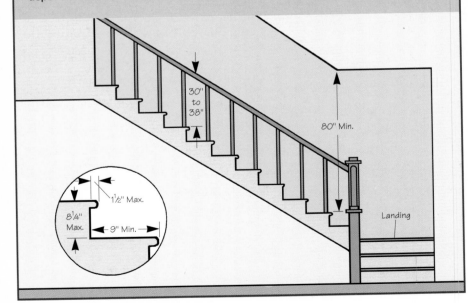

Adding a Balustrade. *For an open balustrade, balusters must be fastened securely to the side of a stringer. Check local building codes to determine proper spacing between balusters; in most cases this is a maximum of 6 in.*

Adding a Partial Wall

Another option is to enclose the stairs on one side with a partial wall that follows the stair pitch. A handrail can be placed on either the partial wall, the full wall, or both. Build the partial wall just as if it were a partition wall with a slanted top plate. Secure the stringer to the studs of the partition wall. Cover both sides of the new wall with drywall or paneling.

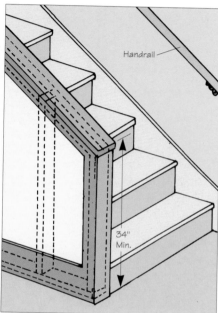

Adding a Partial Wall. *A slanting partition wall can be used to conceal part of the stairs while retaining an open look. You must also install a handrail.*

Attic Stairs

Choosing the location for a stairway to an attic conversion is sometimes an exercise in compromise. Building the stairs, on the other hand, calls for careful attention to details and must never be compromised. An improperly built stairway is a serious safety hazard. If an existing stairway leads to the attic, make sure it's safe enough to withstand frequent use. It may require remodeling or repair.

Calculating the Rise and Run

Once the location for the stairs has been determined, you must calculate the rise of the stairs and the run of the treads and risers. Doing this now allows you to adjust the details somewhat to fit the stairs into the existing space.

Difficulty Level:

Tools and Materials

☐ Tape measure or folding rule
☐ Pencil

1 Figure the Total Rise. The vertical distance between the finished attic floor and the floor below is called the total rise and is key to stairway construction. First measure the distance below the attic; that is, the distance between the ceiling and the floor. If the floor below the attic is carpeted, measure to the subfloor beneath the carpet. Add to this figure the thickness of the existing ceiling and the actual depth of the attic joists. Then add the thickness of the attic subfloor to be installed and the thickness of the floor underlayment, if any. Finally, if the attic will have hardwood flooring, add that thickness, as well. The sum of these measurements is the total distance the stairs rise from floor to floor. Let's say, for example, that the total rise is 102 inches.

2 Calculate the Riser Height. Determine the number of steps that fit into the total rise and exactly how tall each one will measure by dividing the total rise by 7 inches (the ideal riser height as far as safety is concerned). Round down the figure to the nearest whole number. The result is the number of risers needed for the stairs. In the example, 102 inches divided by 7 inches is 14.57. So you need 14 risers. Divide the total rise by the number of risers in order to get the exact height of each riser—this is called the unit rise. In the example, 102 inches divided by 14 equals $7\frac{9}{32}$ inches. For our purposes, $7\frac{1}{4}$ inches will do.

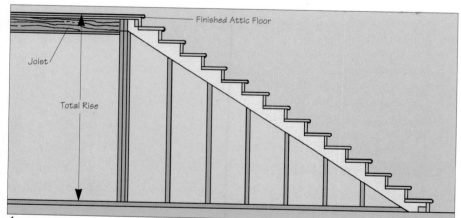

1 *To determine the total rise of the stair, measure from the finished surface of the floor below to the finished surface of the attic floor.*

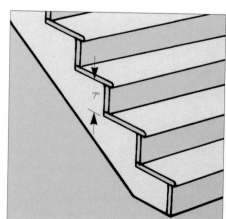

2 *The ideal riser height is 7 in., which offers a good balance of comfort and safety.*

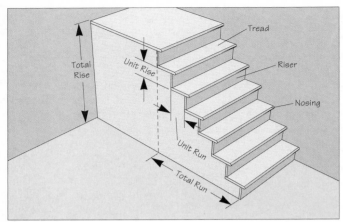

3 To determine the total run of the stairs, multiply the unit run by the number of treads. Note that the actual width of the final tread includes a nosing that's usually 1 in. wide.

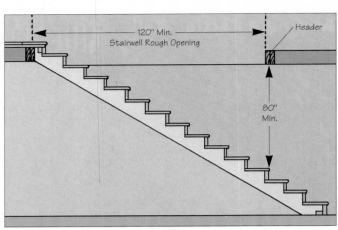

Cutting the Opening. *It helps to make a scaled cross-sectional sketch of the stair area when determining where to begin the rough opening for the stairs.*

3 **Figure the Run.** The number of treads a stairway has is always one less than the number of risers. In this example, you'll have 13 treads. The ideal depth for a tread is 11 inches, but to keep the tread depth and riser height total to 18 inches (see "Basic Stair Dimensions" on page 62), yours would be 10¾ inches. Subtract the depth of the nosing (an inch is common) to arrive at the "unit run" of each step, 9¾ inches. With this figure, it's easy to determine the total run of the stairs (its overall length measured horizontally) by multiplying the unit run by the number of treads. In the example, the total run is 9¾ times 13, or 126¾ inches.

If your space for the total run is unlimited, your calculation is done. You can, however, play with the formula as long as you meet all the tread and riser criteria. Let's say you want to make the total run as short as possible. Try calculating the stairs with the maximum allowable riser height of 8¼ inches: 102 inches total rise divided by 8¼ inches equals 12 risers. That means you need 11 treads. To keep the tread depth and riser height total to 18 inches, you'll need treads that are 9¾ inches deep, including the nosing. Sub-

tracting the nosing leaves you with 8¾ inches times 11 treads, or a total run of 96¼ inches.

Cutting the Opening

To create the stairwell, you must make a hole in the attic floor framing. Though the rough opening can be made perpendicular to the attic floor joists, it's easier if you position the opening parallel to them.

When determining the width of the rough opening, remember to add the thickness of the finished wallcovering. If the finished stairs will be 36 inches wide and the stairwell will be finished with ½-inch drywall, for instance, the rough opening must be 37 inches wide. The length of the rough opening depends on several variables including the run of the stairs. To provide the mandatory 80 inches of headroom at all points on the stairs, however, the opening must be at least 120 inches long.

Building the Stairs

Once you know the run of the stairs, the dimensions of the treads and risers, and the size of the rough opening, you're ready to build the stairs.

Difficulty Level: 🔩 to 🔩🔩🔩

Tools and Materials

☐ Basic carpentry tools
☐ Chalkline
☐ Drywall saw
☐ Framing lumber, stair treads, risers, stringers
☐ Nails, 16d
☐ Joist hangers
☐ Framing square
☐ Circular saw
☐ Drywall
☐ Handrail
☐ Handrail supports

1 **Lay Out the Rough Opening.** Remove all insulation from the area and relocate obstructions such as wiring or plumbing lines. Snap a

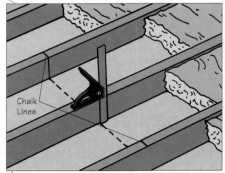

1 *Mark the perimeter of the rough opening and snap a chalkline between these marks. Use a square to locate cut lines on the ceiling. Cover the area below with tarps, then cut into the ceiling.*

chalkline across the tops of the joists to mark the rough opening. Using a square, extend the marks down to the attic side of the ceiling to locate where you'll cut the drywall. Cut through the existing ceiling and remove it.

2 **Cut the Joists.** In most cases, at least two of the existing joists have to be cut to create the rough opening. Use 2x4 braces from below to support the joists temporarily on each end of the opening, then cut through the joists and remove them one by one. To prevent the wood from binding the saw blade, have a helper support each joist from below as you cut.

3 **Frame the Rough Opening.** Trimmer joists should be the same dimension and length as the existing joists. The trimmers help carry the floor load instead of the joists you removed. Install the trimmer joists first, using 16d nails to attach them to the existing joists. Then support the headers and the ends of the cut joists with metal joist hangers.

4 **Frame the Stairwell.** If the stairs will be enclosed, build the walls now. Stairs that are not enclosed must have a free-standing handrail system.

5 **Cut the Stair Stringers.** A stairway is supported by three lengths of framing lumber called stringers. Use good quality 2x12 lumber that's straight and free from loose knots. You'll cut each stringer with a series of identical notches to accommodate treads and risers.

Use a framing square to lay out the notches for the unit rise and unit run. Locate the dimension that corresponds to the unit rise (8 inches in the sample diagram) on the outside edge of the square's tongue. Then locate the unit run on the outside edge of the arm of the square (10 inches in the sample diagram). Step

2 *Have a helper support the joist as you cut it with a reciprocating saw. Shore up any long, unsupported joists with 2x4 bracing.*

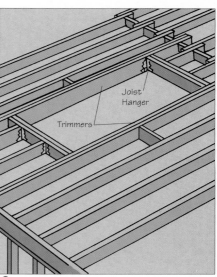

3 *Headers can be attached to trimmers with nails or can be supported on joist hangers. Make sure the trimmers are supported by a wall at each end.*

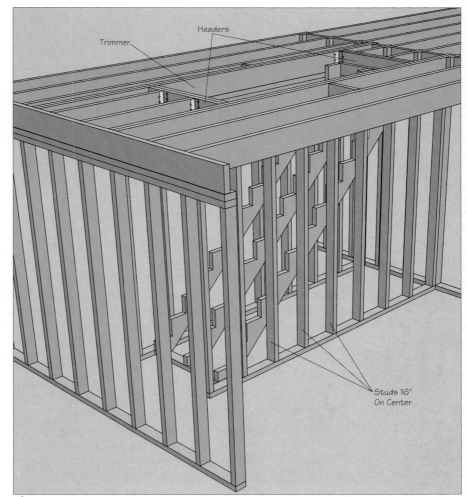

4 *The stairwell walls (if any) are built at this point. Nail the top plate to the underside of the trimmer joists and the bottom plate to the floor.*

off these dimensions along the stringer, then cut out the notches with a circular saw. Don't overcut the notches; use a handsaw to finish off the cuts at each corner.

Once you've cut one stringer, use it as a template for laying out the others. Finally, trim off the top and bottom of each stringer.

6 Adjust the Stringers. The height of the stringers must be adjusted to account for the thickness of the treads. Otherwise, the top step will be shorter than the others and the bottom step will be taller.

This procedure is called "dropping the stringers." Drop the stringers by the amount of one tread thickness by cutting this amount off the bottom of each stringer.

7 Install the Stringers. A kicker made from 2x4 blocking helps keep the stairs in position. Cut a notch in the bottom of each stringer to provide for the kicker. Lift each stringer into place and make sure the treads are level. Space the outer stringers ½ inch away from the surrounding walls so you can slip drywall into place rather than cut it to

match the stair angle. Use a metal framing anchor to secure the top of each stringer, then toenail the bottom into the kicker and nail the kicker to the floor.

8 Install the Treads and Risers. If you plan to carpet the stairs, use 1⅛-inch-thick plywood to make the treads; risers need only be ¾-inch plywood. Use solid wood for stairs that will not be carpeted. If you planned earlier to include a nosing on each tread, add its width to the unit run; the result is the actual depth of the tread. Use a table saw

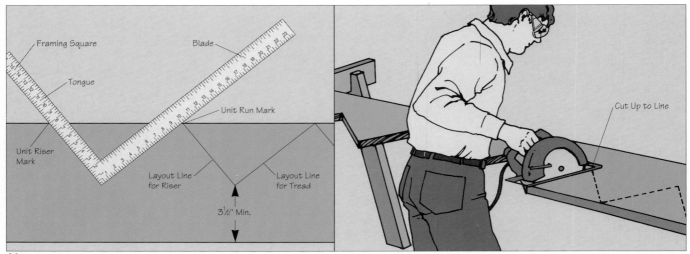

5 *Line up the edges of the stringer with marks on the outside of the square and lay out the unit run and unit rise of the steps onto the stringer (left). Use a circular saw to cut up to, but not through, the layout lines (right). Finish the cuts with a handsaw.*

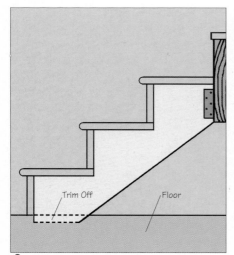

6 *Trim the bottom of each stringer by the thickness of the tread so that the top riser height matches the others.*

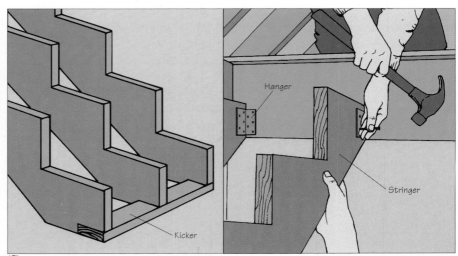

7 *A 2x4 block of wood, called a kicker, helps to keep the bottom of the stairs in position (left). Have an assistant support each stringer as you secure it to the top of the rough opening (right).*

or portable circular saw to cut the risers and treads to size and attach them to the stringers with nails and construction adhesive as you work your way up the stairs.

9 Secure the Handrail. After the interior of the stairwell is finished with drywall, install at least one handrail along the entire length of the stairs, including landings. The handrail must be 30 to 38 inches high. Locate the support brackets along a chalk line and screw each one into a stud. Then cut a handrail to length and attach it to the brackets.

Attic Floors

When the attic or the space above a garage is turned into living space, the ceiling joists for the rooms below act as floor joists for your new room. Ceiling joists are not designed to withstand the loads expected of floor joists, however, so you'll probably have to reinforce them. Don't assume that the existing joists are okay—check them before you proceed with the conversion.

Understanding Floor Loading

All joists are sized to withstand a particular load. In construction, loads are divided into two categories: "Dead" loads, which are static and account for the weight of the building itself, including lumber and finish materials such as tile or drywall, and "live" loads, which are dynamic loads that account for the highly changeable weight of people and furniture. Add the dead loads and the live loads to determine the total load on the joists.

Riser

Tread

8 *Fasten treads and risers with finishing nails. To minimize squeaks, put construction adhesive on the stringers before nailing on the treads and risers.*

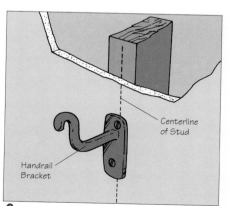

Centerline of Stud

Handrail Bracket

9 *It's important to build a sound handrail. Make sure each bracket is screwed tightly to a stud.*

Building a Platform Stair

If there's not enough space for a straight-run stairway, a platform stairway may solve access problems. This kind of stairway is made up of a pair of short straight-run stairs that are supported by a platform, also called a landing. The platform is built as if it were an elevated section of floor: wood framing supporting joists sheathed with a plywood subfloor. Pay particular attention to codes relating to the platform and the railing.

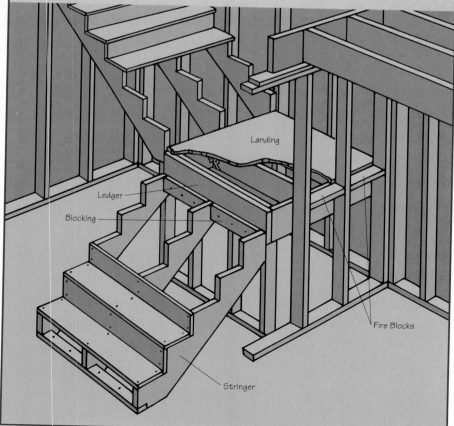

Landing

Ledger

Blocking

Fire Blocks

Stringer

In a home, dead loads usually are figured at 10 pounds per square foot (lbs./sq. ft.) of floor space. Live loads vary. Ceiling joists are designed to withstand live loads as small as 10 lbs./sq. ft. Bedroom floor joists must withstand live loads of 30 lbs./sq. ft., and joists beneath other attic rooms (such as a home office) must withstand live loads of 40 lbs./sq. ft. The ability of a joist floor to withstand a particular combination of live and dead loads is determined by a variety of factors: The species of wood, the depth and thickness of each joist, the spacing of joists, and the distance between joist supports (called the span). All of these factors combined affect not only the strength of the floor but also its stiffness. All floors flex somewhat as they're loaded, though floors with properly sized joists don't flex enough to be noticeable. Before you consult the span tables to size the joists for the proper loads, you must gather some information.

Measuring a Floor for Loading

Difficulty Level:

Tools and Materials

☐ Gloves
☐ Measuring tape or folding rule
☐ Design-value and span tables

1 Measure the Joist Dimensions. Put on protective gloves and push aside the insulation between a pair of joists. Measure the depth of one joist and round the number to the nearest whole number. For example, round 5 ⅝ inches up to 6 inches. Measure the width of the joist and round up that number, as well. The results are the nominal dimensions used in span tables.

2 Figure the Joist Span. The span of a joist is the distance between supports. In most cases,

attic joists rest on the outer wall of the house and an interior wall. Measure the distance between the supports. You may have to push aside more insulation to get an accurate measurement.

3 Measure the Joist Spacing. The closer the joist spacing, the stronger the floor. Measure across the tops of several joists from centerline to centerline. Check in several places around the attic to see if all the joists have approximately the same spacing.

4 Read the Design-Value and Span Tables. Once you've gathered all the information con-

1 *Check the dimensions of existing joists by measuring for depth and width. Round up to the nearest whole number.*

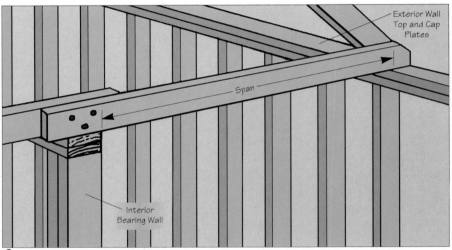

2 *Measure the span of the joist between supports. The result will be shorter than the overall length of the joist.*

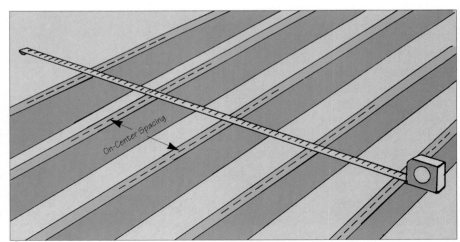

3 *Check the joist spacing in several locations around the attic, as it may vary. Measure from centerline to centerline.*

cerning the structural details of the attic floor, it's time to consult the tables to determine whether or not the existing floor is strong enough to support a living space. Because the tables found in the building code books attempt to account for all the variables that affect floor strength, they can be a little intimidating at first. With a bit of practice, however, you'll see how useful they are.

The first table you need is called "Design Values for Joists and Rafters." It rates the various species of construction woods in terms of their strength (the allowable bending stress, or "Fb") and stiffness (the modulus of elasticity, or "E"). The figure you need is the stiffness, so first find the grade stamp on the lumber used for your floor joist. If the grade reads, "hemlock-fir in a No. 3 grade," for example, then according to the design values table, this wood has a modulus of 1,200,000 (see the table at right).

Take this figure to the tables labeled "Allowable Spans for Floor Joists," which also are found in the code book. Look for the depth of your joists (in this example, 2x6s) on the left side of the chart, then locate their spacing (16 inches on center). Follow the table across until you find figures under the "E" column corresponding to 1,200,000. In this case, the maximum span for these joists is 9 feet 9 inches. If the joists span more than this, they won't support a new bedroom in the attic.

The span table shown here is for sample purposes only and assumes live loads of 30 lbs./sq. ft. and dead loads of 10 lbs./sq. ft. The appropriate table for your project, however, depends on local codes, the species of wood generally used where you live, and the use of the room. If you find that the existing joists are undersized, correct the problem before proceeding with the attic conversion.

Design Values for Joists and Rafters

Species and Grade	Size	Design Value in Bending "Fb" Normal Duration	Snow Loading	7 Day Loading	Modulus of Elasticity, "E"	Grading Rules
Engelmann Spruce–Alpine Fir (Engelmann Spruce–Logdepole Pine) (Surfaced Dry or Surfaced Green)						
Select Structural		1550	1780	1940	1,300,000	
No. 1 & Appearance		1350	1550	1690	1,300,000	
No. 2	2x4	1100	1260	1380	1,100,000	Western
No. 3		600	690	750	1,000,000	Wood
Stud		600	690	750	1,000,000	Products
Construction		800	920	1000	1,000,000	Association
Standard	2x4	450	520	560	1,000,000	
Utility		200	230	250	1,000,000	
Select Structural		1350	1550	1690	1,300,000	
No. 1 & Appearance	2x5	1150	1320	1440	1,300,000	
No. 2	and	950	1090	1190	1,100,000	
No. 3	Wider	550	630	690	1,000,000	
Stud		550	630	690	1,000,000	
Hem–Fir (Surfaced Dry or Surfaced Green)						
Select Structural		1900	2180	2380	1,500,000	
No. 1 & Appearance		1600	1840	2000	1,500,000	
No. 2	2x4	1350	1550	1690	1,400,000	Western
No. 3		725	830	910	1,200,000	Wood
Stud		725	830	910	1,200,000	Products
Construction		975	1120	1220	1,200,000	Association
Standard	2x4	550	630	690	1,200,000	
Utility		250	290	310	1,200,000	West Coast
Select Structural		1650	1900	2060	1,500,000	Lumber
No. 1 & Appearance	2x5	1400	1610	1750	1,500,000	Inspection
No. 2	and	1150	1320	1440	1,400,000	Bureau
No. 3	Wider	675	780	840	1,200,000	
Stud		675	780	840	1,200,000	
Hem–Fir (North) (Surfaced Dry or Surfaced Green)						
Select Structural		1800	2070	2250	1,500,000	
No. 1 & Appearance		1550	1780	1940	1,500,000	
No. 2	2x4	1300	1500	1620	1,400,000	
No. 3		700	800	875	1,200,000	
Stud		700	800	875	1,200,000	
Construction		925	1060	1160	1,200,000	National
Standard	2x4	525	600	660	1,200,000	Lumber
Utility		250	290	310	1,200,000	Grades
Select Structural		1550	1780	1940	1,500,000	Authority
No. 1 & Appearance	2x5	1350	1550	1690	1,500,000	(A Canadian
No. 2	and	1100	1260	1375	1,400,000	Agency)
No. 3	Wider	650	750	810	1,200,000	
Stud		650	750	810	1,200,000	

Allowable Spans for Floor Joists (ft. and in.)

Size (in.)	Joist Spacing (in.)	Modulus of Elasticity, "E," in 1,000,000 psi									
		0.4	0.5	0.6	0.7	0.8	0.9	1.0	1.1	1.2	1.3
	12.0	7-5	8-0	8-6	8-11	9-4	9-9	10-1	10-5	10-9	11-0
	13.7	7-1	7-8	8-2	8-7	8-11	9-4	9-8	10-0	10-3	10-6
2 x 6	16.0	6-9	7-3	7-9	8-2	8-6	8-10	9-2	9-6	9-9	10-0
	19.2	6-4	6-10	7-3	7-8	8-0	8-4	8-8	8-11	9-2	9-5
	24.0	5-11	6-4	6-9	7-1	7-5	7-9	8-0	8-3	8-6	8-9
	32.0					6-9	7-0	7-3	7-6	7-9	7-11
	12.0	9-10	10-7	11-3	11-10	12-4	12-10	13-4	13-9	14-2	14-6
	13.7	9-4	10-1	10-9	11-4	11-10	12-3	12-9	13-2	13-6	13-11
2 x 8	16.0	8-11	9-7	10-2	10-9	11-3	11-8	12-1	12-6	12-10	13-2
	19.2	8-5	9-0	9-7	10-1	10-7	11-0	11-4	11-9	12-1	12-5
	24.0	7-9	8-5	8-11	9-4	9-10	10-2	10-7	10-11	11-3	11-6
	32.0					8-11	9-3	9-7	9-11	10-2	10-6
	12.0	12-6	13-6	14-4	15-1	15-9	16-5	17-0	17-6	18-0	18-6
	13.7	11-11	12-11	13-8	14-5	15-1	15-8	16-3	16-9	17-3	17-9
2 x 10	16.0	11-4	12-3	13-0	13-8	14-4	14-11	15-5	15-11	16-5	16-10
	19.2	10-8	11-6	12-3	12-11	13-6	14-0	14-6	15-0	15-5	15-10
	24.0	9-11	10-8	11-4	11-11	12-6	13-0	13-6	13-11	14-4	14-8
	32.0					11-4	11-10	12-3	12-8	13-0	13-4
	12.0	15-2	16-5	17-5	18-4	19-2	19-11	20-8	21-4	21-11	22-6
	13.7	14-7	15-8	16-8	17-6	18-4	19-1	19-9	20-5	21-0	21-7
2 x 12	16.0	13-10	14-11	15-10	16-8	17-5	18-1	18-9	19-4	19-11	20-6
	19.2	13-0	14-0	14-11	15-8	16-5	17-0	17-8	18-3	18-9	19-3
	24.0	12-1	13-0	13-10	14-7	15-2	15-10	16-5	16-11	17-5	17-11
	32.0					13-10	14-4	14-11	15-4	15-10	16-3

4 *Use tables like these to determine proper joist dimensions.*

Reinforcing a Floor

Reinforcing a floor is not a complicated task, but figuring out the right dimension for reinforcement can be. For this reason, have a structural engineer specify the dimension, spacing, and connection methods for reinforcement. There are several possibilities he or she might suggest. Note, however, that you can't strengthen a joist by simply nailing additional wood on top of it.

Sister Joists. Reinforce an existing joist by joining, or sistering, it to another equal-sized joist. One way to do this is to nail two-by lumber to one side of each existing joist. The new lumber must be as long as the existing joist so that it's supported in the same locations. Slip it into place atop the outer wall plate and cut a small corner off one end to gain clearance beneath the roof.

Add Joists. If the existing joists were closer together, the floor would be stronger. Although you can't readily move the joists, you can add new ones. In effect, this changes the joist spacing of the entire floor. If you don't want to sacrifice headroom in the attic by adding deeper joists, you may be able to add enough joists of the same dimension as the old ones. To determine whether or not this

Add Joists. *Make the floor in the attic stronger by toenailing additional joists to the wall plates.*

works for your situation, check the joist span tables under several different joist spacings.

Stiffen Joists. If the existing floor system is strong but not stiff enough for the job, an engineer might recommend that you attach strips of plywood to each side of each joist. Plywood is only 8 feet long, however, so it won't span the same distance as the joist—that's why it's useful for stiffening the system but not for strengthening it. Nail and glue the plywood to each joist.

Checking the Subflooring

The subfloor has two important roles: It stiffens the floor system and serves as a base for the finished floor. The proper grade and thickness of subflooring material depends on the spacing of the joists and the type of finished flooring to be installed. If subflooring is already in place in the attic, make sure it's suitable. Subflooring made from 1x4 tongue-and-groove boards was common before plywood gained popularity. However, a board attic floor usually isn't flat enough to serve directly as a substrate for finished floor surfaces and probably isn't thick enough to handle the live loads of a living space. Covering the boards with plywood will cost you some head-

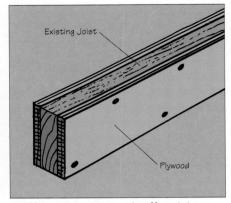

Stiffen Joists. *Plywood stiffens joists, thereby reducing the tendency of the floor to bounce.*

room, so in general it's best to remove them and put them to use elsewhere.

Plywood Types. Plywood is perhaps the most commonly used product for subfloors, though oriented strand board (OSB) is also used. The 4x8-foot size of these panel products covers large areas in a minimal amount of time, and their surfaces are smooth enough for the direct application of some floor coverings without the need for underlayment.

Use an interior-grade subflooring plywood that's thick enough to span between each joist without noticeably deflecting. Plywood is rated for joist spacings of 16 to 48 inches. The greater the spacing, the thicker the plywood. Note that the depth of a joist is not a factor; it is the spacing between the joists that's critical. Subflooring plywood is also graded by use. Panels suitable for use directly under carpeting and pad are in one category, while panels that require underlayment before the application of floor coverings (particularly vinyl) are in another category.

Lastly, subflooring plywood is available in two edge configurations: square edge and tongue-and-groove edge. You can use either kind of edge under any kind of flooring, but

Sister Joists. *Reinforce an existing joist by attaching another equal-sized joist.*

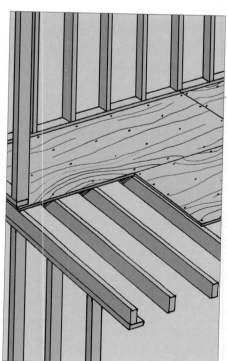

Plywood Types. *Square-edged plywood (top) is more common and less expensive than tongue-in-groove plywood (bottom), which eliminates the need for blocking beneath the seams.*

the cheapest, easiest choice depends on what you're planning for a finished floor.

Square-edged plywood is easier to put down and usually cheaper than tongue-and-groove. However, unless you'll be covering the plywood subfloor with underlayment that's at least ½ inch thick or wood strip flooring that's at least ¾ inch thick, you'll have to install blocking between the joists so that all edges are continually supported. With the underlayment, you must stagger the joints so they don't fall directly over the same joints as those on the plywood subfloor.

If you're not planning a ¾-inch-thick wood floor over your subfloor, you'll probably find tongue-and-groove cheaper and more convenient because it eliminates the need for blocking between joists even if you don't use underlayment. As a result, installing tongue-and-groove boards designed to be used without underlayment is a great way to go if you're planning to use resilient flooring, tile, or carpet.

Installing Subflooring

Lay the subfloor far enough into the eaves so that it supports the knee-walls that may be installed later. There's no need to run the subfloor all the way to the wall plates unless you plan to use the space behind the kneewalls for storage. Subflooring material is usually nailed to the joists. For a stiffer floor that's less likely to squeak later on, run a bead of construction adhesive along the top of each joist before nailing the plywood.

NOTE: Sometimes the hammering that goes along with nailing causes fragile plaster ceilings below the attic to crack; it may also cause drywall compound joints to crack or break loose or drywall nails to pop. To minimize the risk if you have plaster ceilings, don't nail the subflooring material. Instead, use an electric variable-speed drill outfitted with a Phillips screwdriver tip to install the subfloor with drywall-type screws called deck screws.

Difficulty Level:

Tools and Materials

☐ Basic carpentry tools
☐ Plywood, ¾-inch
☐ Nails or screws
☐ Variable-speed drill with Phillips bit

1 Plan the Work. Always lay plywood with the best face up and the grain of each panel running at right angles to the joists. This makes the best use of the plywood's strength. Also, start every other row of plywood with a half-panel so that end joints will not run continuously along the floor. Make sure each edge of every sheet rests on a joist. Add blocking, if necessary, between the joists. The ends of all panels must be centered on a joist.

2 Lay the Panels. Lay panels from the centerline of the floor toward the eaves so that you have a solid platform from which to work. When laying a subfloor over joists that overlap a supporting wall, install blocking between the overlaps. This

1 *Install subflooring with staggered seams and the long dimension running across the joists.*

2 *To avoid back strain, do not pick up the panel. Lift one edge and slide it into place.*

supports one edge of the plywood. Offset adjacent sheets as needed so that edges fall over the joists.

3 Nail or Screw Down the Panels. Manufacturers recommend that you allow a gap of ⅛ inch between square-edged panels to allow for expansion. You can gauge this distance easily by using spacer nails: Slip an 8d nail between panels as you position them. Nail the panels with 6d cement-coated nails or screw them down with 1 ¾-inch deck screws in two corners and then check the position of the sheet before securing it down completely. Place nails at 6-inch intervals along edges of the sheet and at 10-inch intervals elsewhere.

Basement Floors

A smooth, unblemished floor is an asset to any basement remodeling project. Cracks must be repaired even if you plan to install an insulated subfloor and particularly if you plan to paint the floor. An insulated subfloor will help to keep the room cozy.

Repairing Cracks in a Concrete Floor

It's rare to find a concrete floor slab that's completely free of cracks and damage. Minor cracking and small areas of damage are handled easily, though larger areas may call for partial removal of the slab and the advice of a contractor. In any case, repairs must be made before a subfloor or a finished floor is added. Always use safety goggles and gloves when removing concrete or working with patching products.

You can patch cracks up to ⅜ inch wide with hydraulic cement. If you have wider cracks, you probably have serious foundation problems and

3 *Spacer nails create the required ⅛-in. gap between the plywood sheets. Drive nails every 6 in. at the edges and every 10 in. in the field.*

should consult a foundation contractor. Hydraulic cement sets in 3 to 5 minutes, so mix up only what you can use in that amount of time and clean up overspread material promptly. It's wise to wear gloves when handling cement.

Difficulty Level: 🔨

Tools and Materials

- ☐ Goggles
- ☐ Gloves
- ☐ Hand-drilling hammer
- ☐ Cold chisels (various sizes)
- ☐ Wire brush
- ☐ Shop vacuum
- ☐ Hydraulic cement
- ☐ Trowel

1 Clear the Crack. If the crack is wide enough that the blade of a cold chisel fits into it, open it up and undercut the sides so that the beveled edges anchor the patch. A cold chisel, made of less brittle steel than a wood chisel, is designed for masonry work. Use a wire brush to remove loose debris from the edges of the undercut area, then vacuum out the debris and dust. Use hydraulic cement to make the patch.

2 Prepare the Crack. Before using some brands of hydraulic cement, you must soak the area to be patched with water to keep the concrete from wicking moisture away from the patch. Once the

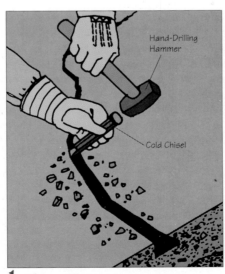

1 *Use a cold chisel to enlarge the crack in the concrete slab. Undercut the crack slightly to provide a beveled "key" for the patching compound.*

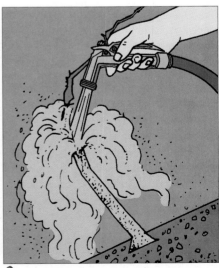

2 *Flush the crack with water to prevent dry concrete from wicking moisture away from the patching cement. Remove traces of dust and debris.*

standing water has been absorbed by the concrete, proceed with the patch. Follow instructions found on the product's label.

3 **Fill the Crack.** Mix a batch of hydraulic cement, and force it into the damaged area. Use the edge of a trowel to tamp the material into place, then use the flat surface of the trowel to level and smooth the patch.

3 *Use the edge of the trowel to pack the patching cement into the crack, then smooth the cement flush with the surrounding surface.*

Floor Overlays

If the surface of the slab has minor damage over a wide area, or if the surface is too rough to serve as a finished floor, you can top it with a new surface. Overlay compound is a gypsum-based liquid product that's self-leveling. You pour it over the floor to a thickness of up to 1/2 inch and spread it with a floor squeegee. The floor, when fully cured, is smooth and uniform. Follow instructions on the product label. Note that it's necessary to contain the product as it's being installed to keep it from flowing into drains.

Painting a Concrete Floor

Paint is an excellent choice for those who want to keep remodeling costs way down or if the basement simply will be used as a workshop. A painted floor prevents stains from reaching the concrete itself, making them easier to clean. Paint also seals the surface against dusting, a powdery residue that sometimes forms on the surface of concrete. A properly prepared concrete surface, along with the right kind of paint, ensures success. Make sure the paint you use is designed for use on concrete floors.

Concrete gains most of its strength soon after being poured but continues to cure for years afterward. If the house is new, let the concrete cure for at least two years before painting it. Make sure moisture problems found in an older house are solved before the floor is painted.

Use an etching liquid to rough up a smooth, shiny surface. This product contains a mild acid, so follow the manufacturer's application and

safety instructions to the letter. Rubber gloves and eye protection are mandatory. If the concrete already feels slightly rough to the touch, use trisodium phosphate (TSP) or a phosphate-free cleaner to clean stains and heavily soiled areas. Then vacuum the floor to remove dust. Pour modest amounts of paint directly onto the floor, then use a medium-nap roller to spread the paint. The idea is to keep the leading edge of the painted area wet so the roller strokes blend together. Spread the paint evenly, otherwise it won't dry properly. Apply a second coat of paint after the first is dry—usually within hours.

Installing an Insulated Subfloor

Most likely there's no insulation beneath your basement slab. One exception might be an unfinished daylight basement, where at least one wall is exposed to the grade. An uninsulated slab is uncomfortably cold in the winter, so contact the builder, if possible, to determine whether insulation exists.

Painting a Concrete Floor. *Pour a modest amount of paint directly onto the floor and spread it with a roller. Keep the leading edge of the painted area wet so the roller strokes blend together.*

An insulated subfloor installed over a concrete slab isolates the finished floor from the slab, resulting in a warmer floor and helping to prevent moisture from damaging the finished flooring. The insulated subfloor consists of plywood that's nailed to sleepers, usually pressure-treated 2x4s laid flat. Rigid insulation fits between the sleepers. Either square-edge or tongue-and-groove-edge plywood can be used, but tongue-and-groove plywood eliminates the need for blocking placed beneath unsupported plywood edges. Remember, there must be at least 90 inches of headroom (84 inches in kitchens, hallways, and bathrooms) after the insulated subfloor has been installed.

Difficulty Level:

Tools and Materials

- ☐ Basic carpentry tools
- ☐ Sheets of 6-mil polyethylene
- ☐ Caulking gun
- ☐ Construction adhesive
- ☐ Masonry nails
- ☐ Pressure-treated 2x4s
- ☐ Extruded polystyrene foam panels
- ☐ Plywood, ¾-inch
- ☐ Cement-coated 6d box nails

1 **Put Down a Vapor Barrier.** After sweeping the floor slab, cover it with sheets of 6-mil polyethylene plastic. Overlap each seam by at least 6 inches. Lift up the edges of the polyethylene and use a caulking gun to put down dabs of construction adhesive to hold it in place.

2 **Install the Perimeter Sleepers.** Use 2 ¼-inch-long masonry nails to nail pressure-treated 2x4 sleepers around the perimeter of the room. If the lumber is dry and straight, a nail or two installed every several feet will suffice. Mark these perimeter sleepers for additional sleepers 24 inches on center. This spacing is suitable for ¾-inch plywood.

1 *Sweep the slab and lay 6-mil plastic sheeting as a vapor barrier, overlapping the seams by 6 in.*

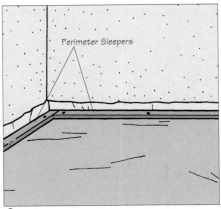

2 *Nail 2x4 pressure-treated lumber around the perimeter of the room on top of the plastic vapor barrier.*

3 *Space pressure-treated 2x4 sleepers at intervals of 24 in. on center. The sleepers provide bearing for the plywood.*

4 *Cut 1 ½-in. thick extruded-foam panels to fit between the sleepers and lay them in place.*

3 **Install Interior Sleepers.** Align the interior sleepers square to the marks on the perimeter sleepers. Use one nail at the end of each board and one about every 48 inches. If one sleeper doesn't extend completely across the room, butt two sleepers end to end.

4 **Insert the Foam Panels.** Medium-density extruded polystyrene foam is best for concrete floor slabs. Use a thickness that matches the thickness of the sleepers, about 1 ½ inches. Cut the pieces to fit between the sleepers and insert them.

5 **Attach the Subfloor.** Use ¾-inch plywood subflooring, either square or tongue-and-groove

5 *Attach ¾-in. plywood panels perpendicular to the sleepers, staggering the joints. Use 6d cement-coated box nails to secure the plywood.*

edge. Cut the plywood to span across rather than parallel to the sleepers. Lay the panels in a staggered pattern.

Garage Floors

Garages are built on foundations just like houses. Except at the garage door opening, a garage will have a small concrete perimeter foundation wall extending about 6 to 10 inches above the slab (more for those houses on steeper grades). Garage floor slabs are poured after foundations have been poured and cured. This enables concrete finishers to install sloped slabs so water will run toward the garage door opening when it drips off of rain-soaked or snow-covered cars parked in the garage. A foundation wall equal in height to the existing foundation must be poured across the garage door opening. The new foundation wall at the door opening is not needed for bearing support, as the existing header above serves that purpose. It needs to be installed as a means of keeping water from coming in contact with wall framing and siding members and from seeping into the garage conversion.

Closing the Foundation Wall

First, the garage door must be removed and all door jamb material removed. This should expose an open end of the garage wall foundation at each side. The foundation wall at this point should be between about 6 and 10 inches tall, or the height of a 2x6, 2x8, or 2x10.

Difficulty Level: ⚒⚒

Tools and Materials

- ☐ Goggles
- ☐ Gloves
- ☐ Basic carpentry tools
- ☐ Hammer drill and masonry bits
- ☐ Small sledge hammer
- ☐ Rebar, ½-inch-diameter
- ☐ Form lumber
- ☐ Ready-mix concrete

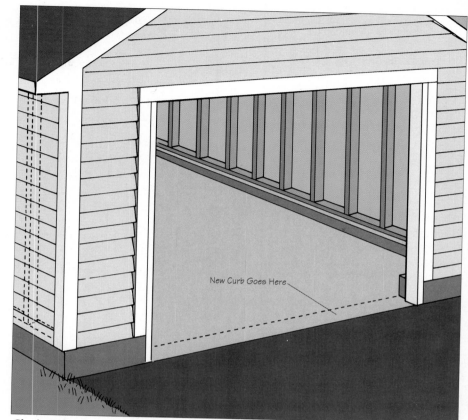

Closing the Foundation Wall. *The opening left behind after the removal of a garage door needs to be "curbed." This is not so much for load bearing (as the wall will be non-bearing) but more for keeping water out of the conversion.*

- ☐ Wheelbarrow
- ☐ Shovel
- ☐ Water
- ☐ Trowel
- ☐ J-bolts

1 Remove the Door Jambs. With a pry bar and hammer, remove the door jamb material to expose the existing foundation wall. Wear safety glasses and gloves.

2 Install Rebar. Use a hammer drill and a ½-inch bit to bore two horizontal holes 6 inches deep into the ends of each of the exposed concrete foundation walls. The holes should be centered and spaced equally from the top and bottom of the wall. Use a small sledge hammer to drive 12-inch pieces of ½-inch rebar into each hole (6 inches into the existing foundation and 6 inches exposed). The rebar will help anchor

1 *Use a pry bar and a hammer to loosen and remove the jambs around the garage door opening. This will expose the studs used to frame the garage and the foundation upon which they rest.*

the new foundation to the existing walls.

3 Secure the Forms. Lay two 2x6s across the opening, one flush with the inside face of the existing foundation and the other flush with the outside face of the foundation. Bolt the ends of the forms to the existing foundation with masonry anchors. Use heavy blocks, rocks, or bricks to hold the forms in the middle. Concrete is heavy and may force weak forms to move outward. If need be, use long 2x4s instead of blocks to brace the forms against walls, trees, curbs, and the like.

4 Pour the Concrete. Mix bags of ready-mix concrete with water in a wheelbarrow, then shovel it into the forms. Be sure to surround the rebar completely. Use a trowel to smooth concrete off flush with the tops of the forms.

5 Install J-Bolts. Use J-bolts to secure the bottom plate of the new wall that will replace the garage door. Install the J-bolts at each end of the form, approximately 6 to 12 inches from the ends of the existing foundation walls, and at about 36-inch intervals. Be certain that you don't insert J-bolts where studs will

be located by measuring out from one foundation end across the opening at 16-inch intervals (the spacing for studs).

NOTE: Each bag of ready-mix concrete comes complete with instructions as to how much water to add and how many cubic feet the bag will fill. Determine the cubic feet of concrete needed inside the forms and buy bags accordingly. Foundations are typically 6 inches wide. Therefore, a foundation wall 6 inches wide by 6 inches deep and 16 feet wide (typical width for a double garage door) would equal 4 cubic feet.

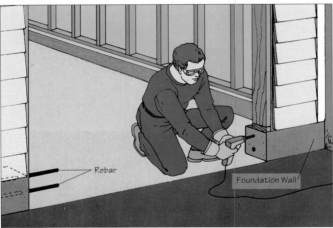

2 Use a hammer drill to bore holes in the sides of the existing garage foundation so you can insert 12-in. pieces of $^1/_2$- or $^5/_8$-in. rebar (depending upon the local code) into them. Be sure at least 6 in. of rebar sticks out into the opening.

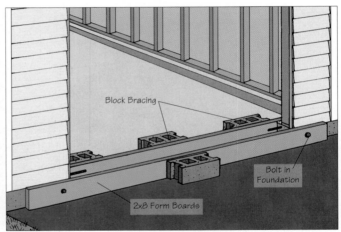

3 Use 2x6 or 2x8 lumber to form a trough at least 6 in. wide. Bolt the forms in place. Use concrete blocks to stabilize the center sections of the forms. The resulting curb should match the existing foundation wall at the sides of the opening.

4 Mix bags of ready-mix concrete in a wheelbarrow and shovel it into the forms. Be sure to surround the rebar completely. Go slow so you don't cause the forms to bow out.

5 After you've filled the forms with concrete, install J-bolts. Determine beforehand where studs will be placed, then plan to locate J-bolts between them.

Garage Floor Options

With few exceptions, home garages will have a concrete floor in fairly good condition. Concrete garage floors in severely cracked or crumbling condition must be replaced. Depending on the headroom available and the condition of the damaged slab, a new 4-inch-thick concrete slab may be poured over the top of the damaged slab or a new wood flooring system may be built over the old slab. In some cases, it may be best to dig up and remove the old slab and replace it with a new one. Consult a concrete or flooring professional to help you arrive at a reasonable and cost-effective solution. The same applies to dirt floors typically found in older garages that were originally constructed as detached buildings.

Newer homes with attached garages will have a concrete floor that should be in acceptable condition. However, this floor will most likely slope toward the garage door opening by about $\frac{1}{8}$ inch per foot. Although hardly noticeable for a workshop or possibly a recreation room, a garage floor with a significant slope could be awkward and uncomfortable for home offices, bedrooms, and other living-space options. One way to level out a sloping garage floor is to pour a new flat and level slab over the top of the existing one. If the first floor of your house is set slightly higher than the existing garage floor, this may be the easiest and least expensive option. Consult a concrete contractor.

Another option is to install a sleeper and plywood floor system like that for a basement (see "Installing an Insulated Subfloor," page 73). The only difference is that you'd use shims under selected sleepers to raise the sloped section of the garage floor high enough to match the sleepers at the highest part of the slab. The new floor will only be raised a few inches, but it will be level and you'll be able to insulate it.

Building an Elevated Subfloor

A third option is to install an elevated subfloor. Houses with wood subfloors, as opposed to concrete slabs, may have a set of two or more steps that lead from the house floor to the concrete garage floor. Usually, a house such as this is built out of the ground on a foundation with a crawl space or basement beneath, and the garage floor is a concrete slab poured on grade. If the roof over the garage is at the same level as the roof over the house, the garage will have a lot of headroom, generally in excess of 8 feet. To make such a garage conversion blend best with the existing house, consider building a subfloor on top of the garage floor slab at a height equal to the house floor. You'll use joists and, depending on the size of the room, a support girder.

Difficulty Level: 🔩🔩

Tools and Materials

- ☐ Basic carpentry tools
- ☐ Water level
- ☐ Framing lumber
- ☐ Sheets of 6-mil polyethylene
- ☐ Stapler
- ☐ Common nails, 16d
- ☐ Joist hangers
- ☐ Common nails, 10d
- ☐ Unfaced fiberglass insulation
- ☐ Plywood, $\frac{3}{4}$-inch
- ☐ Cement-coated 6d box nails

1 **Draw a Guide Line.** Use a water level to mark a level point around the perimeter of the garage. Mark the wall, or the studs if there's no finished wall, every few feet. Measure up or down from one of the level marks to the desired height of the floor, minus the subfloor and finished floor dimensions. Cut a block of wood to this same dimension, then use the block to mark off the floor height around the perimeter of the room. Connect the marks with a long spirit level. The resulting line will be the reference line for nailing perimeter framing to the wall.

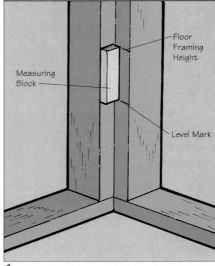

1 *Mark studs at a level line around the perimeter of the garage using a water level (top). Measure down or up from the marks to the desired height of the new floor framing, cut a block to that dimension, and use the block to measure off lines on the walls (bottom).*

2 Nail the Ledgers. Determine the size joists you'll use for the room by consulting the span tables from "Measuring a Floor for Loading," page 69. Lay a vapor barrier of 6-mil-thick plastic over the concrete floor. Overlap the seams by about 6 inches, and run the plastic up the walls to about the reference line you marked before. Staple the plastic to the studs, then nail ledgers, or rim joists and header joists, around the perimeter of the room aligned with the guide line. Use 16d common nails. If the span of the room is too great or you just want to shorten it to use smaller joists, build a girder by nailing together three layers of joist material with 10d nails, staggering the joints. Set the girder at joist height and toenail it to wall studs with 16d nails before installing the rim joists. Support the girder with blocking.

3 Install the Joists. Nail joist hangers spaced 16 inches on center to the header joists and the girder if you use one, then set the joists into the hangers and secure them with 8d or 10d nails.

4 Install Blocking. Nail blocking between the joists every 48 inches or where plywood edges will fall as bearing for nailing, unless you're using tongue-and-groove plywood subflooring.

5 Insulate and Lay Subflooring. Install unfaced fiberglass insulation (R-19 for 2x6s, higher for larger-dimension lumber) between the joists and flush with their tops. Nail or screw down ¾-inch plywood subflooring with the long edge perpendicular to the joists. Use 6d coated box nails or 1½-inch deck screws every 6 inches on the edges and every 10 inches elsewhere. To help prevent squeaking or popping, run a bead of construction adhesive along the tops of the joists. The floor structure is ready for finished flooring.

2 Lay plastic on the slab as a vapor barrier. Determine the size joists you need, then nail rim joists and header joists of the same size around the perimeter of the garage, using the line marked in Step 1 as a guide (left). If the span is too great or you want to use smaller joists, make a girder from three joists to run mid-span (right).

3 Install joists 16 in. on center using hangers on the headers and girder.

4 Nail two-by blocking between each joist every 48 in.

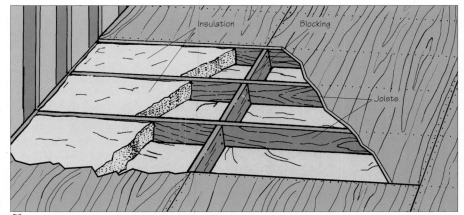

5 Insert unfaced fiberglass insulation between and even with the tops of the joists. Secure ¾-in. plywood panels perpendicular with the joists, using nails or screws.

Attic Framing and Dormers

5

Almost all attic conversions call for the construction of new walls and ceilings. Whether the walls are full-height partitions or half-height knee-walls, framing them properly makes it easier to install the finished wall surfaces later in the project.

Building Partition Walls

A partition wall extends to the ceiling, dividing the attic space. It does not, however, play a role in the structural integrity of the house. In an attic, partition walls are used to enclose a bathroom or to divide the attic into separate rooms. Usually the walls used in an attic aren't large, so they can be built one at a time on the subfloor and tipped into place. If space is limited, you may have to piece the wall together in place. In the places where a partition wall meets a sloped ceiling, you must cut a series of angled studs to fit between the wall plates and follow the ceiling slope.

Difficulty Level: 🔩🔩 to 🔩🔩🔩

Tools and Materials

☐ Basic carpentry tools
☐ Framing square
☐ Chalkline
☐ Framing lumber
☐ Circular or power miter saw
☐ Nails, 10d, 12d, 16d
☐ Level, 48-inch

Using the Tip-Up Method

Most partition walls are built with 2x4 lumber and have a single top and bottom plate. If a wall is to contain the drain line for plumbing fixtures, however, you might want to use 2x6s to frame it. All ceiling framing or collar ties must be in place before you frame the walls. There's usually not enough room to slip them into place afterward.

1 **Mark the Location.** Use a framing square to ensure square corners and a chalkline to mark the exact location of the wall on the subfloor. If you're working alone, drive a nail partway into the subfloor to hold one end of the chalkline as you snap it. Or if you have one, get a helper to hold one end.

2 **Lay Out the Plates.** Cut two plates so they're the same length as the wall. Align the plates and measure from one end, marking for studs at 16-inch intervals on center, the standard spacing for studs. Continue to the other end of the plates even if the last stud is less than 16 inches from the end. To check your work, mark a point exactly 48 inches from the end of

the plates. If done correctly, the mark will be centered on one of the stud locations.

3 **Cut the Studs.** Cut wall studs to the height of the wall less twice the thickness of the lumber to account for the thickness of the plates. Count the number of layout marks on the plates to get an estimate of the number of studs you'll need. Then cut the studs square.

4 **Build the Frame.** Separate the plates by the length of a stud and set them on edge with the layout marks facing each other. Lay all the studs in their approximate positions, then drive a pair of 16d nails through each plate into the ends of each stud. Use the marks to align the studs precisely.

5 **Form the Corners.** To provide a nailing surface for the drywall, add an extra stud to each end of the wall that's part of an outside corner. One method of building the corner involves nailing spacers between two studs, then butting the end stud of the adjacent wall to this triple-width assembly. Another method is to use a stud to form the inside corner of the wall. Use whichever method you find most convenient.

1 Lay out the position of the walls using a framing square and a chalkline. Mark an X on the side of the line that will be covered by the plate.

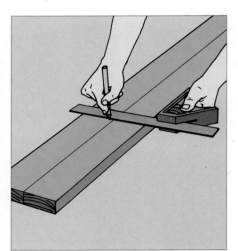

2 Use a combination square to mark the position of each stud on the plates. By marking the plates simultaneously, you can be sure the layouts match.

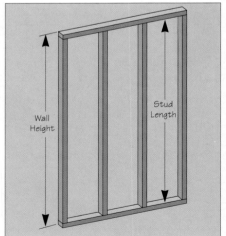

3 Use a circular saw to cut each stud to fit exactly between the plates. Stud length is the wall height minus 3 in., or the depth of two two-bys.

4 Using 16d nails, attach the studs and plates, keeping them aligned and flush. Blunted nail tips won't split the plate.

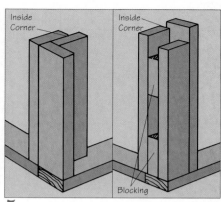

5 Inside corners must have a nailing surface for installing drywall.

6 When raising a wall, tip it fully upright, then slide it into position according to layout marks.

7 Use a level to plumb the wall. Check several places on the wall as you nail the top plate.

6 **Raise the Wall.** Slide the bottom plate into approximate position according to the subfloor layout lines. Then lift the wall upright. With a helper to prevent the wall from toppling over, align the bottom plate with the layout lines made in Step 1. When you're satisfied with the location of the wall, nail a pair of 16d nails every 24 inches or so, through the bottom plate into the subfloor (into joists whenever possible). If the ceiling in the room below is made of plaster, use long screws rather than nails to fasten the plate to prevent the ceiling from cracking.

7 **Plumb the Wall.** Use a 48-inch level to ensure that the wall is plumb. Adjust the frame as necessary and nail through the top plate into the ceiling framing or into blocking if the wall runs parallel to the ceiling framing.

8 **Join Intersecting Walls.** You'll need additional studs to provide support for drywall in the places where walls intersect. Add a single stud to the end of the intersecting wall and a pair of studs on the other wall.

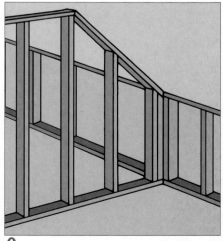

8 As you join walls, consider how the drywall will be installed. Add extra framing or blocking, if necessary.

Building a Wall in Place

When there's not enough room to assemble a complete stud wall in the confines of the attic subfloor, you must build the wall in place. You'll slip each stud between the top and bottom plates.

1 Install the Top Plate. Cut both plates to the length of the wall and mark them for the positions of the studs at 16 inches on center. Determine the location of the top of the wall and hold it there, making sure the stud layout faces down so you can see it. Then use a 16d nail to attach the top plate to each intersecting rafter.

1 When nailing the top plate to the ceiling joists, set the nailheads flush with the surface of the plate. Otherwise, you may have trouble fitting the studs.

2 Locate the Bottom Plate. To transfer the location of the plate to the subfloor, hang a plumb bob from the top plate in several successive locations, marking them as you go. Align the bottom plate with the layout marks, adjusting it until you're sure that it's directly below the top plate. Then use pairs of 16d nails to nail the plate to the subfloor. Keep the nails away from the stud locations already marked on the plate.

3 Install the Studs. Measure between the plates at each stud location and cut studs to fit. Place a stud in position against the layout marks and use 12d nails to toenail it to each plate. Make toenailing a bit easier by using a spacer block to keep the stud from shifting. Cut the block to fit exactly between studs. If the stud spacing is 16 inches on center and the studs are 1½ inches thick, the block will need to be 14½ inches long. Remove the spacer as you toenail successive studs.

2 Use a plumb bob to position the plate, then nail it to the subfloor. Transfer stud locations from the top plate to the bottom plate in the same way.

Building a Sloped Wall

Some walls have one or two top plates that match the angle of the rafters. If ceiling joists were installed for a flat ceiling and you want the wall to go all the way across the attic, for example, there will be three top plates: One under the ceiling joists and one under each sloping section of the wall. In any case, a sloping wall can be built in place much the same as a partition wall. The major difference is that the studs vary in length and have angled cuts at the top.

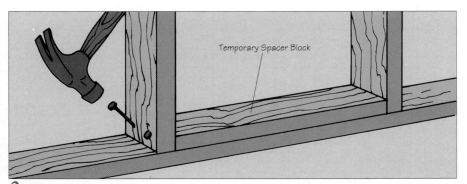

3 A stud may shift off the layout lines as you toenail it to the plate. Hold the studs in place with a temporary spacer block. Make sure the nail doesn't hit the block.

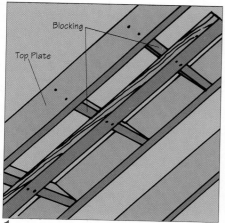

1 *Blocking provides a nailing surface for top plates located between rafters.*

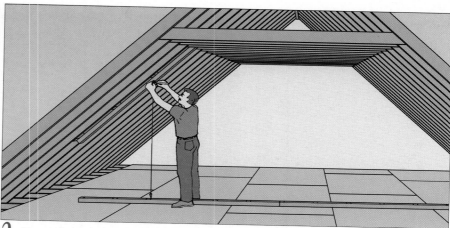

2 *Use a plumb bob to transfer the stud layout from the bottom plate to the top plate.*

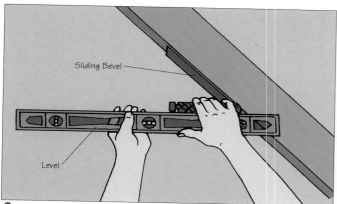

3 *Place the sliding bevel against the rafter, level the square, and tighten the blade.*

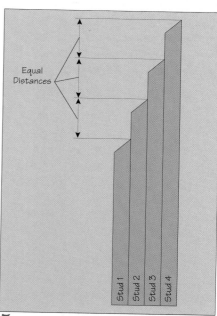

4 *Set the circular saw's blade angle with the sliding bevel. Hold the wood securely, and slowly make the cut.*

1 **Install the Plates.** If the wall is situated between rafters, install 2x4 blocking between the rafters to provide a nailing surface for the top plate of the wall. There must always be at least two blocks for every wall; longer walls may need blocks on 24-inch centers. If blocking is installed, snap a chalk line across the blocks to mark the position of the top plate. Nail the top plate to the blocks or to the bottom of a rafter. Drop a plumb bob at both ends of the top plate to mark the position of the bottom plate. Nail the bottom plate to the floor.

2 **Mark the Stud Locations.** Lay out the studs 16 inches on center across the bottom plate. Then use a plumb bob to transfer these locations to the top plate. Don't try to lay out the positions by measuring along the angled top plate; if you do, the studs won't be 16 inches on center.

3 **Capture the Angle.** To replicate the angle of the rafters, use an adjustable sliding bevel placed against a level.

4 **Cut the First Stud.** Measure the distance between plates to get the length of the first stud (measuring to the "high" side of the angle). Then set the angle on a circular saw and cut across the face of the stud.

5 **Cut Successive Studs.** Measure and cut the second stud just as you did the first. Then hold the two against each other and measure the difference in length between them. You can use this measurement,

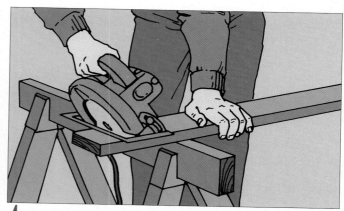

5 *Use the difference in length between the first two studs to determine the other stud lengths.*

6 *Nail into the top of a stud until the nails just poke through the ends, then put the stud into position and complete the nailing.*

Kneewalls

Kneewalls are the walls that extend from the floor of an attic to the underside of the rafters. They're not tall: 48 inches is a common height because it matches the width of a drywall sheet. The 2x4 studs typically are spaced 16 inches on center. In most cases kneewalls are not structural; the rafters continue to carry the roof loads. A kneewall can be adapted easily during the framing stage to support drawers, cabinets, and other objects that make use of space in the eaves. Another good idea is to build in some sort of access to the spaces behind the walls. This allows you to do routine maintenance and to install additional electrical outlets if ever the need arises.

called the common difference, to determine the length of all remaining studs that are spaced the same distance apart. Each one is longer than the one before it by the amount of the common difference.

6 **Install the Studs.** Use a pair of 10d nails at the top and three 10d or 12d nails at the bottom of each stud to toenail them into place. Use a spacer block or your foot to keep the bottom of a stud from shifting as you nail it.

Installing Doors and Door Framing

The framing around attic doors is fairly simple to build because the partitions are not load bearing and there's no need for a structural header above the door. Many do-it-yourselfers find prehung doors easiest to install. Prehungs can be installed whether or not a structural header is in place. These factory-assembled units eliminate some fussy carpentry work. Because the size of the rough opening depends on the size of the door and its frame, however, purchase the door before you frame the wall. The rough open-

ing is generally ½ inch wider and ¼ inch taller than the outside dimensions of the jamb.

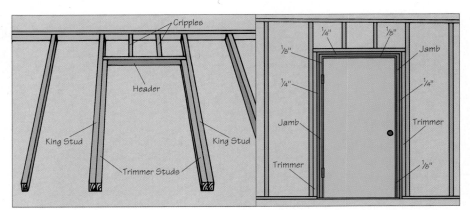

Installing Doors and Door Framing. *In a wall that contains a door, be sure to account for king studs and trimmers on the bottom plate. Clearances around the door (rough opening) are found in the manufacturer's instructions.*

Kneewalls. *You can build a shelving unit in a non-structural kneewall. The unit is a plywood case that fits into the rough opening. The drawers fit between studs.*

Installing Structural Kneewalls

If the wall has to support sagging rafters it may call for a structural kneewall. In this case, the wall must be designed not only to carry the loads but to transfer them properly to a part of the house that can bear them. You glue and nail plywood to one or both sides of the kneewall, turning it, in effect, into a giant beam. Sometimes the bottom plate is doubled to distribute loads over the floor system. But because the details of structural kneewalls must be worked out with care, it's best to let a structural engineer design them. The fee for such work is generally modest. Drawings and calculations that the engineer provides become a part of your application for a building permit.

Building Non-Structural Kneewalls

A non-structural kneewall has a single top plate and a single bottom plate. The stud spacing may be varied somewhat, but remember that the wall has to support drywall, so you'll need a stud for each vertical joint. There are several ways to construct kneewalls.

Angled Plate. Some builders prefer to rip an angle along the edge of the top plate to provide plenty of support for the drywall. The angle is best cut on a table saw. Use 2x6 stock; otherwise, the plate will not fully cap the studs. Assemble the wall, then tip it into place and secure it with 16d nails. To make nailing easier, adjust the stud layout so you can nail through the plate into each rafter.

Standard Plate. A kneewall that has a 2x4 top plate is easier to build than one that has an angled plate. After assembling the wall, tip it into place. Use wood shims, if necessary, to make sure the wall is tight against the rafters, then nail through the plate and into the rafters with 16d nails. Secure the bottom plate with 16d nails.

Drywall Nailers. Both of the methods above result in a solid kneewall that has plenty of blocking for finished wall surfaces. You must also provide blocking to support the ceiling edges, however. The best way to do this is to cut 2x4 blocks to fit between the rafters. Measure each space between rafters to account for small differences in the spacing. Attach one end of each block by face-nailing through a rafter into the block

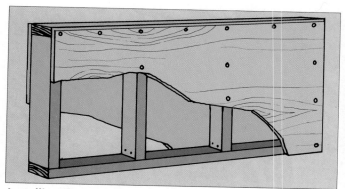

Installing Structural Kneewalls. If a kneewall must support undersized rafters or other loads, have it designed by an engineer. You can create a strong beam by nailing on plywood.

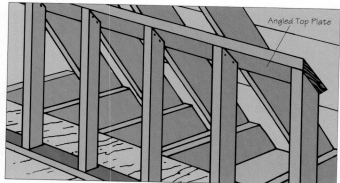

Angled Plate. The top plate of a kneewall must be ripped from 2x6 stock for the front edges of the studs and plates to be in the same plane.

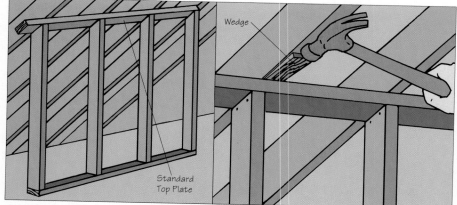

Standard Plate. A wall with a standard 2x4 top plate (left) is easier to build than one with an angled plate. Use wood wedges to fill the gaps between the plate and the rafters. Make the wedges snug (right), but don't push the rafter out of position.

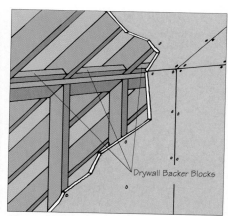

Drywall Nailers. Use scraps of 2x4 nailed to the rafters to "back" the drywall where it meets the top of the kneewall.

with 16d nails. The other end won't be accessible to face-nailing, so toe-nail it to the rafter with 8d nails.

Insulation and Ventilation

It's important to allow at least 2 inches of clear space above ceiling insulation. This space allows the moisture that migrates through the insulation to exhaust through roof vents. In hot weather, the ventilation also helps keep the underside of the roof sheathing cooler.

Installing Rafter Vents

Some insulation may exist between the attic floor joists, or the house's upper ceiling joists. If you already have insulation here, look for it near the eaves of the roof. Often the insulation found there blocks the air coming in through soffit vents. To ensure proper ventilation, install channel rafter vents in the lower portion of the rafter bays.

Ventilating the Roof

The ideal strategy for roof ventilation is to draw in air through soffit vents and exhaust it through a continuous ridge vent. Other vent combinations, such as gable vents and soffit vents, can be used as long as an ample amount of air flows over the insulation. A ridge vent is the only effective option for a vaulted ceiling.

Installing Insulation

The recommended amount of insulation is expressed as an "R-value." The higher the R-value, the more effective the insulation. Check with local building officials to determine the amount of recommended insu-

lation for your area. In much of the Northeastern United States, for example, R-38 is recommended for roofs. Amounts vary considerably depending on the severity of the climate. Install insulation meticulously because gaps "short circuit" the thermal envelope and encourage drafts.

CAUTION: *When working with fiberglass insulation, be sure to protect yourself from the fibers that inevitably get released into the air. Wear a dust mask, eye protection, a long-sleeved shirt, a hat, gloves, and long pants when cutting or moving insulation.*

Existing Insulation. If what will be the attic floor is insulated, you don't have to remove the insulation as long

as it doesn't protrude more than a few inches above the level of the joists. Insulation here is actually good because it reduces some sound transmission through the attic floor. By leaving it in place you also avoid the messy job of removing it.

Wall and Ceiling Insulation. Properly installed insulation keeps an attic living space comfortable throughout the year. Various products can be used, but fiberglass batts are the easiest to install, particularly when it comes to the sloped rafter bays of the ceiling. Some people insulate the entire roof system, running insulation all the way from ridge to eave. This is often done when kneewalls contain built-in storage behind which it would be difficult to insulate. Other

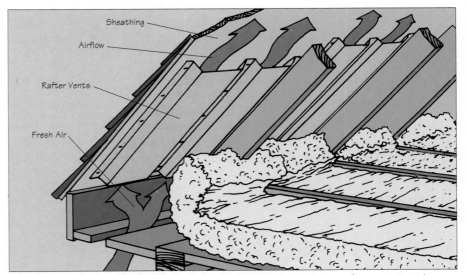

Installing Rafter Vents. *Staple lightweight plastic channels, or rafter vents, to the roof sheathing to ensure that air can pass from the soffit into the rafter bays.*

Ventilating the Roof. *Ridge and soffit vents provide airflow beneath the roof sheathing. Nail shingles to a ridge vent to make it unobtrusive.*

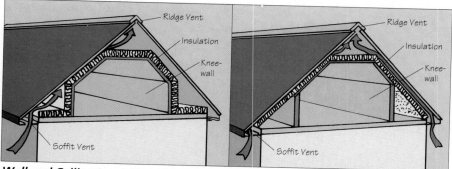

Wall and Ceiling Insulation. *A continuous layer of insulation in the ceiling and knee walls ensures a tight thermal envelope (left). Insulate the rafter spaces all the way to the outside wall plate (right) when you use knee walls for storage.*

people opt for the smallest possible thermal envelope by insulating the kneewalls and just part of the ceiling.

Insulating an Attic Roof

Difficulty Level:

Tools and Materials

☐ Insulation baffles
☐ Stapler (½- and ¼-inch staples)
☐ Fiberglass batt insulation
☐ Sheets of 6-mil plastic
☐ Contractor's tape
☐ Caulking gun and caulk
☐ Rigid foam insulation
☐ Strapping, 2x3
☐ Screws (2½-inch) or 12d nails
☐ Drywall T-square or straightedge
☐ Duct tape

When you plan to use the attic for living space, you'll have to insulate at least part of the roof. The cavities between rafters may or may not be deep enough to contain the desired thickness of insulation while still allowing 2 inches above for air to circulate under the roof deck. If more insulation is needed, you can add rigid foam below the cavity insulation. The following steps show how to insulate between the rafters. Be sure that all rough wiring and electrical junction boxes are in place before beginning the insulation work.

1 Install the Insulation Baffles. Fitting blanket insulation into the spaces between rafters in a way that leaves a clear 2-inch airspace under the roof deck is a hit-or-miss proposition unless you first install insulation

baffles against the roof sheathing. Made of rigid plastic, these devices are formed into a corrugated shape that guarantees air channels next to the roof. Insulation baffles are available at most home centers and lumberyards in lengths of 48 inches and widths to match 16- and 24-inch-on-center rafter spacing. Simply place each baffle in the cavity and use ½-inch staples to attach it to the roof deck.

2 Secure the Batts. Batts faced with kraft paper or foil are available in widths of 15 or 23 inches. To install the insulation, you staple the tabs that extend past the width of the batts along the rafters. Use

1 *Fasten insulation baffles to the roof deck with ½-in. staples.*

2 *Staple faced insulation in place through paper flanges (left). Secure unfaced insulation with tiger's teeth (center) or string (right).*

¼-inch staples spaced every 6 inches. If the rafters are closer than standard width, cut batts along their length and create tabs at the cut ends by rolling back the insulation an inch or so.

If you're using unfaced batts, secure them with wire supports called tiger's teeth. The supports are slightly longer than the cavity width so they bow upward, pressing against the batts and digging the ends of the tiger's teeth into the sides of the rafters. Tiger's teeth are available only for rafters spaced 16 or 24 inches on center. To support unfaced batts between other spacings, staple string in a zigzag pattern across the bottom of the rafters. If you're work- ing alone, staple sections of about 24 to 48 inches at a time, then insert a batt at one end and pull it through the section. The process is faster when a helper holds each batt in position while you string and staple the portion below it.

3 **Install the Vapor Barrier.** A separate vapor barrier is recom- mended for unfaced batts. Kraft paper offers a marginal vapor barrier but only if extreme care is taken to seal all joints and overlaps with duct tape and/or caulk. The foil on foil- faced insulation can serve as a vapor barrier when the joints are sealed. To ensure a continuous vapor barrier regardless of the type of insulation,

3 Install a continuous layer of polyeth- ylene as a foolproof vapor barrier.

staple a sheet of 6-mil polyethylene over the face of the rafters. Over- lap the seams and tape them with contractor's tape. Use tape and/or flexible caulk (like polybutylene or silicone) to seal the polyethylene to abutting surfaces, such as end walls, floors, and the like.

Adding Rigid Foam

You may find that the rafters in your attic aren't deep enough to house the recommended amount of insu- lation. If this is the case, consider adding rigid foam insulation across the underside of the framing. Rigid insulation has a thermal resistance of at least R-4 per inch, making it an effective insulator that doesn't use up a great deal of attic headroom.

Begin by completing all rough elec- trical wiring. Stuff as much blanket insulation as possible between the rafters, being sure to leave an air- space above for ventilation. Then select a type and thickness of rigid foam that makes up the rest of the desired R-value. When choosing a type of rigid foam, be sure to consider cost, the R-value of the board, and whether to use foil facing or a sep- arate vapor barrier. You'll install the foam between runs of wooden strap- ping placed across the rafters. Try to match the thickness of the foam to the strapping to be used. Standard thicknesses of strapping are ¾ inch for 1x3s and 1½ inches for 2x2s, 2x3s, and 2x4s.

1 **Mark the Lines for Strapping.** Measure along one sloped side of the ceiling, marking off intervals of 24 inches. Repeat this process at the other side. Have a helper hold one end of a chalkline while you snap a line across the bottoms of the rafters.

2 **Attach the Strapping.** Use 2x3s or rip 2x4s down the center to make two pieces of about the same size as a 2x2. Starting at the top of the ceiling, place a piece of strapping

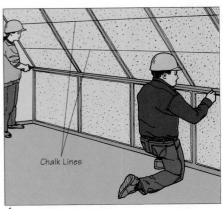

1 Snap horizontal chalk lines at 24-in.- on-center intervals.

2 Attach strapping with 2½-in. drywall screws or 12d nails.

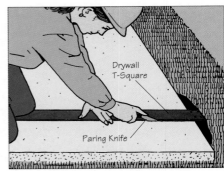

3 Cut the rigid foam with a paring knife and straightedge.

so that its top edge is on the chalk line. Then use 2½-inch-long drywall screws or 12d box nails to attach the strapping to the rafter. If you use screws, predrill the strapping pieces before inserting them.

3 **Cut the Foam.** Lay a sheet of rigid foam insulation over a cutting surface, such as scrap ply- wood or carpet. Measure and mark pieces to fit between the strapping

4 *Fit and secure each cut piece to the strapping with tape.*

pieces (22½ inches for 2x2s). Using a drywall T-square or other straight-edge as a guide, cut through the foam. Use a paring knife to make two or three passes, cleanly cutting through the foam and facing material.

4 **Insert the Foam.** Fit each piece of cut foam snugly between the strapping. Cut smaller pieces, as necessary, to trim out the edges. Put pieces of duct tape over the corners and where necessary to hold the cut pieces in place until you install the wall and ceiling finish materials. When all foam is in place, install a polyethylene vapor barrier, followed by the finish materials.

Dormers

Building a dormer may seem like a small project, but it actually calls for a diverse collection of construction skills—many of the same skills necessary to build an entire house. Still, building a dormer or two is well worth the effort for those who desire additional natural light, ventilation, and a way to make the most of existing floor space.

Types of Dormers

Dormers change the roofline of the house and provide headroom where it's needed most (near the eaves). Because they're visible outside the house, you'll want the dormers to complement the style of the house. Use the same roofing, siding, and window type and style used on the rest of the house. There are three basic types of dormers: Gable dormers (sometimes called doghouse dormers), shed dormers, and eyebrow dormers.

Gable Dormers. A gable dormer has a roof with two pitched planes that meet at a ridge. The ceiling is usually vaulted. The roof pitch need not match that of the house, but it may help the dormer fit in better visually. A valley at the intersection of the

dormer roof and the surrounding roof channels away water. A gable dormer is good for creating natural light and ventilation, but its proportions restrict its size, making it ineffective at increasing usable floor space.

Shed Dormers. The hallmark of a shed dormer is its single flat roof, which is pitched at less of an angle than that of the existing roof. This kind of roof is easier to build than the roof on a gable dormer. It's also easier to join to the existing roof because shingles simply lap over the intersection. The ceiling inside a shed dormer either follows the upward slope of the rafters or is totally flat. The best feature of a shed dormer, however, is that it dramatically increases the usable floor space in an attic. Some shed dormers even run the length of the house, though the

Gable Dormers. *Depending on its size, a gable dormer accommodates one or more windows. The shape of the dormer ties it visually to the house.*

Shed Dormers. *These are usually larger than gable dormers and are effective at expanding the amount of usable floor space in an attic. A shed dormer can be small enough to house a single window or as long as the entire house.*

magnitude of such a project generally calls for a professional builder or structural engineer. As an option, you may want to add shed dormers on both sides of the roof, an arrangement that resembles saddlebags.

Eyebrow Dormers. Unlike other types of dormers, an eyebrow is used primarily to allow natural light into an attic or to serve as a decorative accent for certain house styles. Its small size and curved shape encourage the use of a fixed, rather than operable, window. The window itself is usually custom-made, but some manufacturers offer a limited selection of stock units.

Other Kinds of Dormers. Sometimes the outer wall of a gable or shed dormer is built in the same plane as the outer wall of the house. The house siding may continue up the face of the gable dormer, or it may be interrupted by a small section of roof supported by short lengths of rafters called stubs or dummies. A dormer designed flush with the siding gains the maximum amount of attic floor space and simplifies framing somewhat but doesn't always fit in with the style of the house.

Another option is to recess the dormer into the roofline. Water-proofing in this case is quite a challenge, so plan to do the work under the guidance of a builder and a roofer. You'll usually find recessed dormers on houses that have a covered front porch.

Eyebrow Dormers. These small dormers are sometimes used strictly as architectural accents, but they can provide light and ventilation as well.

Other Kinds of Dormers. Flush dormers are in line with the front wall of the house. A flush shed dormer (top) dramatically expands the living space. A flush gable dormer (middle) allows the window to be installed almost as low as the attic floor. To maximize the size of the window, tuck the dormer into the roof (bottom).

Planning the Dormer

There are several things to consider when deciding upon the size and style of your dormer. For one, the dormer must be built in proportion to the house; a king-size dormer on a small roof causes the house to appear top-heavy, while a dormer that's too small doesn't admit a worthwhile amount of light. Take a drive around town and observe the various dormers. When you see one you like, note its size compared with that of the house.

Many people end up sizing the dormer to accommodate a particular window size, particularly if the window complements other windows in the house. If this is the case, the dormer must be at least 6 inches wider on each side than the rough opening of the window. Doing this provides strength at the corners of the dormer and accommodates framing details. Check local codes; the dormer window may have to be large enough to serve as an emergency exit, or egress window.

Before you cut a hole in the roof, some careful planning is in order. Draw a section view (cross section, or side view) of the existing attic framing. Then experiment with various dormer designs.

1 **Measure the Slope.** The slope of a roof is traditionally expressed as the number of inches it "rises" for every foot it "runs." Rise is measured vertically, while run is measured horizontally. Use a level and tape measure to figure slope. Mark the level at a point 12 inches from one end, then hold the level against the underside of a rafter until it reads level. Use the tape to measure the distance from the level to the rafter at the 12-inch mark. If the distance from the rafter to the level is 11 inches, for example, the slope is 11 inches of rise in 12 inches of run. This is written as 11/12. Carpenters often express it as "11 in 12."

Another way to measure slope is to place an electronic level against the underside of a rafter; it provides a direct readout of roof slope.

2 **Draw a Cross Section.** Measure the width of the gable end, the outdoor height of its walls, the depth of the rafters, and the depth of the attic floor joists. Use this information, along with the slope of the roof found in Step 1, to draw a cross section of the house to scale. Be sure to include the ridge.

3 **Design the Dormer.** Now that you have a cross section of the house, you can experiment with the size and placement of the dormer. This example shows a shed dormer, but drawing a gable dormer is much the same, except that you might want to draw a front view as well.

Place a sheet of tracing paper over the drawing made in Step 2 so you can make several dormer sketches without redrawing the house. You can also make photocopies of the drawing. Sketch a rafter first, then experiment with various locations for the dormer's face wall. For a shingle roof to drain properly, the rafters must have a slope of at least 3/12. The rafter can extend all the way to the existing ridge, if necessary. Measure from the attic floor to the underside of the dormer rafters to determine the headroom that results. Add details such as the header and plate for the window; this is essentially a cross-sectional view of the window's rough opening. Make sure you include at least a double 2x6 header. The bottom of the rough opening must be at least 6 inches above the plane of the roof to allow room for flashing and window trim.

Once the construction details of the dormer are worked out, you might want to see how it'll look on your house. One way to do this is to make a scale drawing of the front of the house, then draw in the dormer,

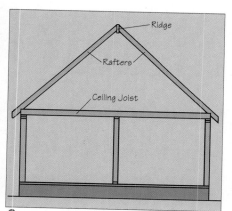

1 *Before building, determine the roof slope with a level and tape measure (top) or an electronic level (bottom).*

2 *A scaled cross-sectional sketch of the house helps to identify the amount of headroom in various places in the attic.*

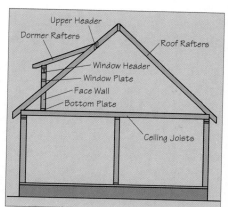

3 *Draw various roof slopes and locations for the face wall. Ask an engineer if the joists can support the dormer.*

along with its window. Another way is to take a photo of the house and use a permanent marker to draw the dormer on it. Take a photo that includes both the front and one gable end of the house (a three-quarter view), allowing you to sketch in both the front and the side of the dormer.

Building a Shed Dormer

You have to cut a large opening in the existing roof to build a dormer, so be sure to have all the necessary tools and materials (including windows) on hand before you start. Being prepared minimizes the amount of time that the house is vulnerable to changes in the weather. Purchase a heavy-duty waterproof tarp to cover the opening overnight or in the event of unexpected rain. Once the dormer is tight to the weather, you can finish the inside. Use fiberglass batts to insulate the walls and ceiling, but don't block airflow above the ceiling.

Building a dormer requires that you spend a lot of time on the main roof, though much of the construction is actually done from inside the attic. Roof work is hazardous. To minimize the risk of injury and to help the work proceed smoothly, be sure to have a stable working platform. A good extension ladder is your best asset. Consider using a ladder hook or roof jacks, which can be rented.

Difficulty Level:

Tools and Materials

☐ Basic carpentry tools
☐ Ladders and accessories
☐ Waterproof tarps
☐ Insulation
☐ Chalkline
☐ Flat spade
☐ Circular saw
☐ Framing lumber
☐ Nails, 10d, 12d, 16d
☐ Bevel square
☐ Framing square
☐ Plywood or OSB panels, ½-inch
☐ Edge, apron, and step flashing
☐ Roofing felt
☐ Shingles
☐ Windows
☐ Siding
☐ Soffit material and soffit vents
☐ Gutters

1 **Locate the Opening.** If you haven't done so already, install plywood subflooring throughout the attic. This provides a safe platform from which to build the dormer and keeps demolition debris out of the floor joist cavities and the insulation, if any. The subfloor must be in place to support the face wall of the dormer.

A shed dormer is built within the confines of a large rectangular hole cut into the roof. Essentially, this hole is the rough opening for the dormer. It's important to locate the rough opening on the sloping underside of the roof properly. First identify the rafters to be removed, then use dimensions taken from the cross-sectional drawing to snap two chalk lines on the attic floor. One line represents the outside of the face wall; the other represents the inside face of the upper header.

2 **Mark the Opening.** Use a plumb bob to determine the points at which the chalked layout lines "intersect" the two rafters (called trimmers) that form the outside of the rough opening. Draw a plumb layout line on the rafters at each intersection, then drive a nail clear through the roof where the layout meets the underside of the roof sheathing. The nails appear on the roof, marking the corners of the opening.

1 *Locate the position of the face wall and the upper header on the subfloor. Both are parallel to the eaves of the house.*

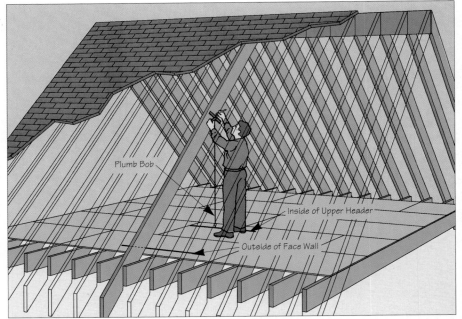

2 *Use a plumb bob to locate the edges of the rough opening and mark them on the rafters. Hammer a small nail through the sheathing to mark each corner.*

3 Set a circular saw to the sheathing thickness. Make the top cut first and stand on the sheathing only while making this cut. Make subsequent cuts from the roof or the plank.

4 Cut the rafters and remove them one by one. If the opening is large, use a temporary framework made of 2x4s nailed into the floor and rafters to brace it.

3 **Strip the Roof and Remove the Sheathing.** Before you start, spread a tarp over the plants and ground below where you're working to prevent nails and such from falling on the shrubs or lawn. After securing the appropriate roof jacks or ladders, snap chalk lines between the protruding nails. Then pound each nail back through the roof to keep from tripping on them later. Use a razor knife to score shingles along the chalk lines, then remove all shingles and roofing paper between the lines. Use a pry bar or flat spade to pry up the shingles.

Set the circular saw to a depth that just cuts through the sheathing. Plywood sheathing is usually ½ inch thick; board sheathing is usually ¾ inch thick. Use a carbide-tipped saw blade designed for demolition work to cut along all four sides of the rough opening. Then from inside the attic, use a hammer and pry bar to remove the sheathing. Pull the sheathing into the attic instead of letting it fall from the roof.

4 **Mark and Cut the Rafters.** Most of the remaining work can be done from inside the attic. At the top of the roof opening, mark each trimmer rafter with a second layout

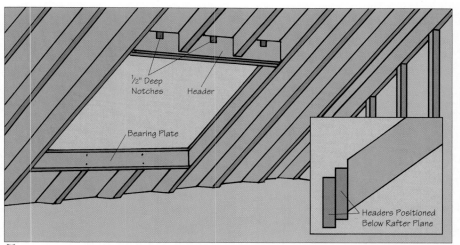

5 Install a header to transfer loads to the rafters on each side of the opening. Use 16d nails to nail through each piece of the header into the ends of the rafters. Do the same with the bearing plate.

line 3 inches from the first—3 inches being the thickness of the header. At the bottom of the roof opening, draw additional layout lines, but make them 1½ inches from the ones drawn in Step 2. Check all lines for plumb.

Use a crosscut handsaw or a reciprocating saw to cut one rafter at a time. Support the rafters with temporary braces before you begin cutting. Make your cuts at the second layout lines to make room for the header above and a "bearing plate," a kind of sill you'll install at the bottom of the opening. Make the

bottom cut first, then have a helper support the rafter as you make the top cut to prevent the wood from binding the saw.

5 **Install the Headers.** Cut three pieces of lumber to fit between the trimmer rafters. This lumber must be the same dimension as the existing rafters (usually 2x8 or 2x10). At the bottom of the opening, fit one of the pieces against the cut ends of the lower rafters, nail it in place, and nail the rafter ends to it. This bearing plate rests against the face wall of the dormer.

The two remaining pieces of lumber become the header at the top of the opening. Before installing the header pieces, cut some shallow notches in the top edges. Two notches in each bay will suffice. The notches allow air to circulate from the dormer's soffit vents into the rafter bays above. Hold up one piece of lumber and nail it in place, then nail through it into the cut ends of the rafters. Slip the second piece into position and nail it in place, then nail through it into the first piece. The two pieces will be offset slightly due to the slope of the roof.

6 Double the Trimmer Rafters.

Each trimmer must be strengthened to carry the additional loads imposed by the header. You can strengthen the trimmers easily by nailing another rafter directly to the outside of each trimmer. This "sister" rafter must be cut to fit exactly between the ridge and the wall plate. Use a sliding bevel to copy angles from the existing trimmers, and nail each sister securely to the trimmer rafter. Once the sisters are in place, nail the cut edges of the roof sheathing to the doubled rafters.

7 Frame the Face Wall. The face

wall of the dormer is usually framed with 2x4 lumber, 16 inches on center. Framing details are the same as those for a regular exterior wall, though you may have to improvise the stud spacing somewhat because of the small size of the wall. Consult the instructions that came with the window to determine the size of the rough opening. Assemble the wall on the attic floor and tilt it into place. Cut one end of the corner posts to match the roof slope, then cut them to length to fit under cap plates. Nail the corner posts to the studs and cap plates, then through the sheathing into the rafters.

8 Cut the Dormer Rafters. There

are several ways to calculate the length of a rafter. You can even get books filled with precalculated rafter tables. Beginners, however, find it easiest to draw a full-size rafter layout. Measure the appropriate dimensions for the header and face wall, then use a framing square and chalkline to draw a cross-sectional view of the header and face wall on the attic floor. Draw in the rafter and use a sliding bevel to copy the plumb cuts. The dormer's end rafters are doubled, and they get a different cut at the top because they land on main

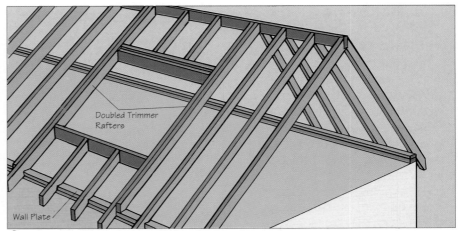

6 Cut two rafters to fit alongside the trimmers and nail them to the trimmers using 16d nails. The tops of the reinforcing rafters fit against the ridge, while the bottoms rest on the wall plate.

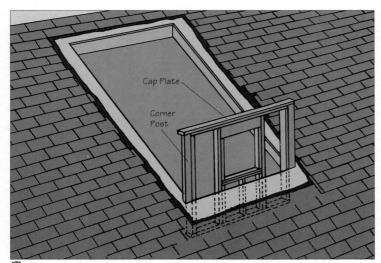

7 Assemble the face wall and tip it into place. Plumb the wall and nail it to the floor and the trimmer rafters with 16d nails. Toenail each stud to the bearing plate. Install the corner posts.

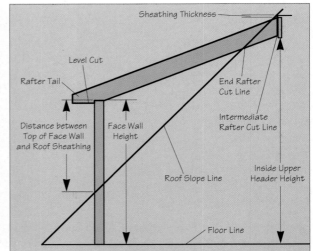

8 Draw a full-scale layout of the dormer rafters and face wall on the attic floor to serve as a template from which to gather measurements for the rafter cuts.

roof rafters instead of meeting the upper header. To find this cut, draw the roof slope line as shown in the drawing. Remember to allow for the sheathing thickness. Then lay out the pattern on rafter stock and make your cuts. Hold the rafter in place to check for fit, and use it as a template for cutting the remaining rafters. Rafters must be 16 inches on center. The rafter tail level cut must be at least 4 inches long to allow room for soffit vents.

9 Frame the Roof and Sidewalls.

Lay each rafter in place, and toenail it first to the header then to the top plate. You can do the nailing without climbing onto the roof. Periodically check the face wall to make sure it remains plumb and straight.

Trim away an additional swath of roofing on each side of the dormer to make room for 2x4 wall plates. Nail each plate through the roof sheathing and into the trimmer rafters, then lay out the locations of the sidewall "studs" so that they're 16 inches on center. Cut the sidewall studs to approximate length, then hold each one in place, mark the angle with a sliding bevel, and cut it to fit between the plate and rafter. Toenail each stud in place, and add blocking if necessary at the top of the sidewall to provide a nailing surface for sheathing.

10 Sheathe and Roof the Dormer.

Use exterior-grade plywood or OSB for sheathing. Apply the wall sheathing first—it stiffens the dormer and makes the installation of roof sheathing less risky. On both surfaces, but particularly on the walls, use large pieces wherever possible. Don't use scraps. This procedure may seem wasteful, but it makes the dormer strong.

After installing rake boards, fascia, edge flashing, and roofing felt, install the shingles just as you would on a roof. Start from the lowest edge and work your way up to where the dormer roof intersects the main roof. Pry up the first course of shingles above the intersection and slip the new shingles beneath them. You may have to adjust the exposure of each shingle course as you install it.

11 Install the Window.

This job is definitely a two- or three-person operation. Lift the window through the rough opening if you can; this is

9 *Use two 16d nails on each side of the rafters to toenail them to the header and wall plate. Angle each piece of sidewall framing, top and bottom, and toenail the studs with four 8d nails.*

10 *Use 6d common nails to nail sheathing every 6 in. at the edges and every 12 in. elsewhere. Use four roofing nails to install shingles over asphalt felt underlayment.*

11 *Nail a flanged window directly to the wall sheathing. An assistant inside the dormer helps plumb and level the window as you hold it in place from the outside.*

12 *Purchase step flashing that matches the exposure of the roof shingles. Slip each piece into place as you replace shingles alongside the dormer. The siding laps the flashing.*

13 *Add and finish trim, gutters, and siding. Don't allow the gable-end wood trim to rest on the roofing at the top of the gable. Cut the trim short so water doesn't soak into the end grain.*

easier and safer than lugging it up a ladder and carrying it across the roof. Have one person hold the window flush against the sheathing while another plumbs and levels the window from inside. Then secure the window to the framing by nailing through the nailing flange.

12 Flash the Walls. The metal flashing installed around dormer walls is much the same as that around a skylight. The metal used may be aluminum, galvanized steel, or another type of metal. Install the apron flashing first, then slip step flashing beneath each course of roof shingles. The work goes slowly when you get to the small space where the two roofs meet, but extra attention here ensures a leak-free dormer.

13 Complete the Exterior. You can install the siding, soffits, soffit vents, and all remaining window trim before painting the dormer. Don't forget to install a rain gutter; water cascading onto the roof below quickly damages shingles. Whenever possible, route the downspout to another gutter. If this isn't possible, put an elbow at the bottom of the downspout and aim the water in the right direction to reduce its impact on the roof shingles.

Building a Gable Dormer

Most of its wall framing is similar to that of a shed dormer, but the multiple roof planes of a gable dormer make it a more difficult project. Cutting and assembling the rafters sometimes seems complicated, but at least you've got the roof framing of your own house to serve as a model. The trickiest part of gable framing arises in places where the dormer meets the surrounding roof—at the valleys. The compound angles at the valleys makes this a project only for those who have advanced carpentry skills.

1 Frame the Walls. Install a subfloor, then locate, prepare, cut, and frame an opening in the roof as for a shed dormer. Use the cross-sectional drawings to make sure you'll have adequate headroom. You can frame the side and face walls of a gable dormer just as you would a shed dormer, or rest the face wall on a header as shown. The dormer ridge projects horizontally from the main roof, with rafters supporting it on each side. Note that the ceiling joists are in the same plane as that of the upper header.

2 Determine the Roof Slope. To cut the rafters to the proper angle and length, you'll have to determine the roof slope you want, which is the number of inches the roof rises per foot of run. The run is the distance from a sidewall to the ridge line. In this example, the distance from the ridge to the building line is 4 feet and the height of the ridge at the top plate is 2 feet, so the rise is 6 inches per foot of run.

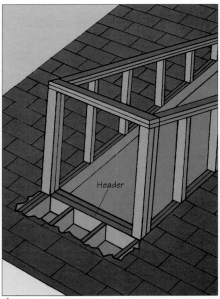

1 *The walls of a gable dormer can be framed like a shed dormer, or the face wall can rest on a header as shown here.*

3 Set the Ridgeboard. Center a 2x4 ridgeboard in the opening so its top is 2 feet above the top plate. Cut the back end of the ridge to fit against the roof. Support the ridgeboard in front with a 2x4 set on end on the front top plate. If you've cut an odd number of house roof rafters, the ridgeboard will meet the center rafter at the header and should be nailed to it. If you've cut an even number of house roof rafters, attach the ridgeboard directly to the header. Level the ridgeboard carefully and nail it with 8d nails. The smaller nails are acceptable because the ridge-board only separates the rafters; it doesn't provide support. Although this ridge isn't a load-bearing part of the structure, it must be correctly centered and leveled.

4 Mark the Ridge Cut. Place the long arm of a framing square along the edge of a rafter board. The short arm, called the tongue, should be on the left, pointing away from you. Pivot the square until the 12-inch mark on the arm and the 6-inch mark on the tongue are aligned with the edge of the board. Draw a line from the top of the board to the bottom, along the tongue of the square. This will create a cutting line that will make the rafter fit against the ridgeboard.

5 Lay Out the Rafters. Mark the 12-inch point where the arm crosses the edge of the board. Slide the square along the edge until the 6-inch point on the tongue aligns with the mark. Repeat sequentially until you reach the building line. Place the 6-inch point on the tongue at the building line mark. Draw a line to the bottom of the board. Go back to the ridge cut line, measure back one-half the thickness of the ridgeboard, and draw a line through the mark.

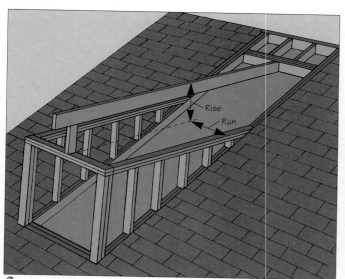

2 Gable dormer roof installation begins with determining the slope of the roof you'll install and attaching a ridgeboard parallel to the existing rafters at the correct height.

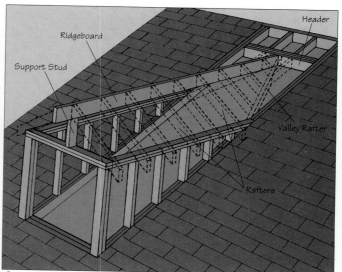

3 Attach the ridgeboard to the support stud in front and the header at the roof. Gable rafters run from the ridgeboard to the sides of the dormer and, at back, to the valley rafters.

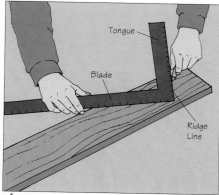

4 Align the blade with the upper edge of the board. Pivot on the 12-in. mark until the 6-in. mark on the tongue intersects the board, and mark it.

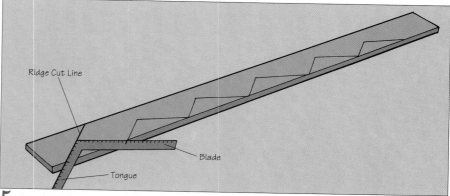

5 To determine the rafter length, lay the framing square along the ridge line with the rise per foot (6 in.) on the tongue and the unit of run (12 in.) on the blade, aligned with the edge. Mark the board where the blade intersects the edge of the board. Lift the square, and starting at your last mark, repeat for each foot in the run.

6 Lay Out the Bird's Mouth Cuts. Now reverse the position of the square so that the tongue is on the right and points toward you. Align the inside of the tongue with the building line and position the square as before, aligning the 6- and 12-inch marks on the board. Draw a line along the inside of the square along the arm and the tongue. This outline marks the "bird's mouth" you'll cut so the rafter can fit over the top plate. Now slide the framing square back toward the top edge and align the 12-inch mark of the arm with the building line and the 6-inch mark of the tongue on the upper edge. Draw a line down the tongue to mark the end of a 1-foot overhang. Cut the rafters and toe-nail them in place with 10d nails.

7 Build the Overhang. Notch the end rafters to accept 2x4 lookouts, or blocking, which will act as nailing for the flying rafter. Set the blocking in place, nail through the common rafter where the blocking butts it, and nail down through the blocking into the notched end rafter. When the blocking is installed, sheathe the roof with plywood.

8 Finish the Gable Dormer. The dormer essentially is a miniature version of a standard roof and is shingled the same way. Take particular care where the dormer roof intersects the main roof. Metal flashing beneath the shingles protects both valleys.

The gable dormer will have a gable-end overhang and a soffit on each side. Finish the overhang, soffits, and fascias as you would for a main roof.

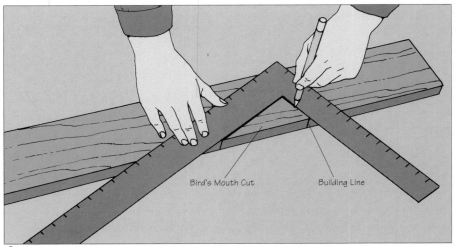

6 *Mark the point where the rafter reaches the wall; align the inside of the tongue with the mark to draw the bird's mouth.*

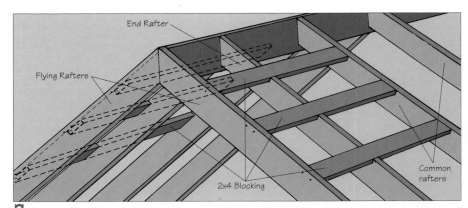

7 *Notch the end rafter to accept 2x4s that serve as lookouts, or blocking, for the flying rafter. The rafter and blocking create an overhang and soffit to shade the window from direct summertime sunlight.*

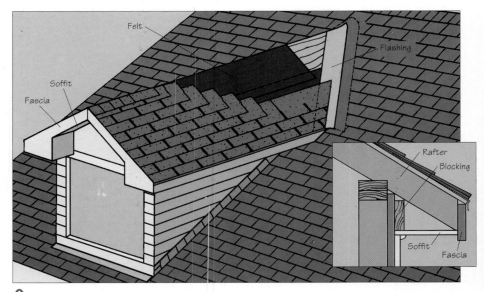

8 *Build the soffits and fascias as you would for a main roof. Cover the plywood roof sheathing with 15-pound roofing felt, fitting it under the metal valley flashing. Start the shingles at the bottom and work your way up to the ridge.*

Basement and Garage Framing

6

C onverting a basement or garage to living space is different from remodeling other areas of the house because you must do a lot of work involving masonry walls and concrete floors. With a few specialized tools and materials, however, you can build basement or garage walls and insulate a new room with ease.

Fastening Objects to Masonry

In the course of most basement or garage remodeling projects, you'll be faced with fastening objects to a masonry surface. You might have to install shelf-support brackets or furring strips on concrete-block walls or anchor the bottom plate of a partition wall to a concrete floor. The task of fastening objects to masonry calls for special tools, fasteners, and techniques. The process may also call for a protective dust mask because it sometimes creates an abundance of fine, abrasive dust.

Hammer Drill. A power drill is indispensable when it comes to drilling into masonry. Consider buying or renting a hammer drill, preferably one equipped with variable speed, if you have to drill many holes. This tool creates a hammering motion as it spins the bit. The dual action helps shatter aggregate in the concrete and clear dust from the hole. Hammer drills often come with built-in depth gauges.

Drill Bits. You can recognize a bit designed for use in masonry by its enlarged carbide tip. Though more brittle than steel, carbide holds up well to the abrasive process of drilling into masonry. The flutes help to clear dust and debris away from the hole. Bits with "fast-spiral" flutes are the ones most commonly seen in hardware stores and home centers, but they're not suitable for drilling through wood and into masonry in one pass (as when setting wall plates). In this case, use a masonry bit with regular "twist" flutes. Some masonry bits have reduced shanks that fit into standard 3/8-inch drills.

Not all masonry bits can be used in a hammer drill, however. The percussive action can damage or destroy some grades of carbide. Look for a note on the packaging to make sure the masonry bit you'll use in the hammer drill is approved specifically for this use. In other cases, some hammer drills will accept only a particular style of hammer-drill bit, referred to as "SDS" bits. These bits are designed to slide into the special chuck on some hammer drills. When the chuck is rotated to the locking position, the bit stays in place and is ready for drilling.

Choosing Masonry Fasteners

At one time, the choice for securing anything to masonry was limited to lead anchors. These days, however, there are many products from which to choose. Nearly all of them fall into one of two broad categories: mechanical anchors that grip the masonry or chemical anchors that bond to it. For most fastening jobs encountered in the course of a basement or garage remodeling project, mechanical anchors work satisfactorily and are generally less expensive and more widely available than chemical anchors.

Lead Sleeve Anchors. The kind of masonry anchor that most people know (and few love) is called a lead anchor. It consists of a lag screw and a lead sleeve, or shield, that fits into a hole that's drilled in the masonry. When you turn the bolt into the sleeve, the sleeve expands and grips the sides of the hole. These fasteners are readily available and inexpensive, but their holding power is limited. Another disadvantage is that two holes have to be drilled: a fairly large hole in the masonry for the anchor and a smaller one in the piece to be fastened for the lag screw. Use lead

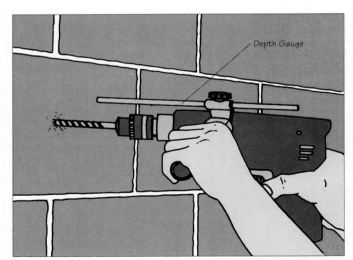

Hammer Drill. *A variable-speed drill equipped with a masonry bit will bore holes in concrete. However, a hammer drill with a depth gauge will do the job faster and more effectively.*

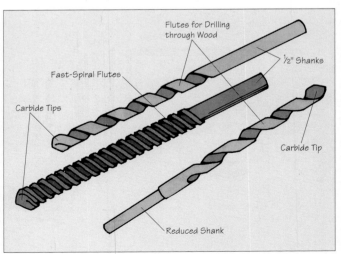

Drill Bits. *The flutes show whether a bit can be used for wood (top) or only for masonry (middle). A reduced shank lets the bit fit standard 3/8-in. drills (bottom).*

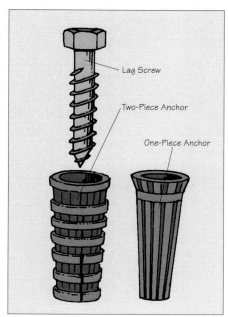

Lead Sleeve Anchors. Lead sleeves are available in one- and two-piece versions.

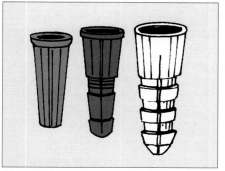

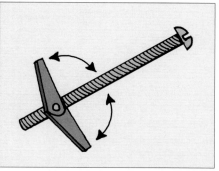

Plastic Sleeve Anchors. These are slightly tapered and hold better than lead sleeves.

Hollow-Wall Anchors. The "wings" of the anchor fold flat so that the assembly can be inserted through a small hole.

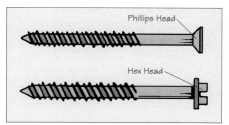

Concrete Screws. Masonry screws have two sets of threads and a Phillips or hex head.

anchors to install shelf cleats or in other situations where the screws are not subjected to a lot of pulling force, particularly if only a handful of fasteners are needed.

Plastic Sleeve Anchors. A better anchor has a plastic sleeve instead of a lead sleeve. You must still drill two different-sized holes, one in the masonry and one in the workpiece, but the hole you drill in the concrete for the plastic sleeve is usually smaller than the one required for a lead anchor, a distinction that makes a big difference if there are many anchors to install. A more important advantage, however, is that plastic anchors hold better than lead anchors. That holding power makes them a better choice for hanging heavy objects, such as cabinetry, from a masonry surface.

When using either plastic or lead anchors, the hole you drill must be just slightly deeper than the length of the anchor. If it's too shallow, the anchor can't seat properly and will not hold as well.

Hollow-Wall Anchors. When working with concrete block it's usually

best to drill right into the solid web of the block. This area offers the greatest holding power for sleeve-type fasteners. It may not be possible to guess where the web is located, however, or a fastener may be needed in a place where there's no web. In such cases it's best to use a hollow-wall anchor, sometimes called a toggle bolt. The most common version features a set of spring-loaded "wings" that are threaded to fit around a bolt. After the wings are slipped through a hole, they expand and grip the back-side of the hole as you tighten the bolt. These fasteners are inexpensive but can be awkward to use. Once the wings are in the hole, the bolt can't be withdrawn without loosing the wings inside the wall.

Concrete Screws. Though it seems implausible, certain types of specially hardened screws can be driven directly into concrete or concrete block without requiring a sleeve. These screws, often referred to by the brand name Tapcon, actually cut threads in the masonry. After drilling a pilot hole, simply turn the screw into the hole as if it were a wood screw. Though they're relatively expensive and not universally available, concrete screws are well worth the effort it takes to obtain them. They hold as well as most sleeve-type anchors and don't require two separately sized holes.

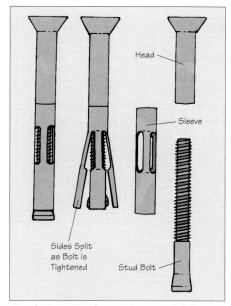

Stud-Type Anchors. You need drill only one hole for this kind of anchor. The sides of the sleeve are split so, when you tighten the bolt, they spread and grip the sides of the hole.

Stud-Type Anchors. Another kind of fastener that's designed for concrete, sometimes called by the tradename Thunderbolt, consists of two parts. One part is a threaded bolt. The other part is shaped like a stud with split sides. You drill one hole through the piece to be fastened and into the concrete. Then tap the two-piece fastener into the hole with a hammer. After it's seated in the hole, you turn the bolthead with a screwdriver. As you tighten the bolt, the stud shaft expands and wedges itself in the hole. This is an excellent fastener for securing furring strips to concrete walls and plates to concrete floors. You need drill only one hole about ¼ to ½ inch in diameter, depending on the size of the fastener.

Masonry Nails. For speedy installation, nothing beats a masonry nail. These fasteners have a thicker shank and heavier head than standard nails of the same length; they also have flutes that lock into masonry. The holding power of masonry nails isn't great, but because it's so easy to install them, it's possible to use more of them. The nails are particularly

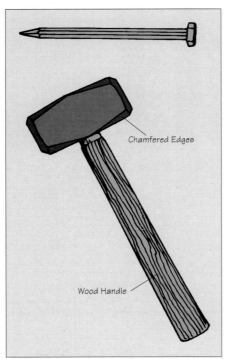

Masonry Nails. *The nails are hardened to resist bending. The extra heft and special head of the hand-drilling hammer (bottom) makes it the best (and safest) for striking masonry nails.*

useful when securing wood cleats to a wall and when fastening wall plates to a concrete floor. Masonry nails can also be used on concrete-block walls, but you should nail them into the mortar joint rather than into the blocks themselves. Concrete blocks are brittle and tend to fracture and crumble when subjected to the stress caused by masonry nails.

Because masonry nails are treated with heat in a special process called "hardening," they don't bend as you drive them into masonry. Hardening, however, means that you must never drive the nails with a standard framing hammer. The nails are likely to damage the hammer head, which can send razor-sharp metal shards into the air. Use a hand-drilling hammer instead.

CAUTION: When working with masonry nails, wear eye protection to shield yourself from flying masonry chips.

Chemical Anchors. Though mechanical anchors handle most fastening tasks, special situations require the use of a chemical anchor—where the hole has to be made at the edge of a concrete block, for example. While most mechanical anchors stress the masonry around a hole and are more likely to cause the masonry to crumble, chemical anchors do not add stress.

The various chemical products work in similar ways: A hole is drilled in the masonry and filled with an epoxy-like mixture, then a threaded rod is pushed into the hole and held in place until the adhesive sets. Some adhesives can be mixed thick enough to resist dripping when used on a vertical surface. Chemical anchors, though relatively expensive, tenaciously grip the masonry. They're usually available at larger hardware and home-center stores.

Building Partition Walls

The wide-open space of a full basement or a garage is perfect for a pool table or play area, but if the area is going to be used for something else, it may be better subdivided into smaller spaces with partition walls. In most cases it's possible to build the walls one at a time on the basement or garage floor, then "tip" them into place. Where space is limited or where obstructions such as pipes or beams make it hard to tip up a wall, you can assemble partition walls in place.

If you plan to run a partition wall perpendicular to the joists, all you have to do to anchor the top of the wall is to nail through the plate and into the joists using 16d nails. If you'll run the wall parallel to the joists, however, you'll have to install 2x4 blocking between the joists to provide a nailing area. Blocking is usually set 16 inches on center, but in this case you'll offset the blocks from the partition studs to make it possible to nail into them through the top plate with a pair of 16d nails.

Difficulty Level: 🔨🔨 *to* 🔨🔨🔨

Tools and Materials

☐ Basic carpentry tools
☐ Framing square
☐ Chalkline
☐ Framing lumber
☐ Common nails, 10d, 12d, 16d
☐ Spirit level, 48-inch
☐ Masonry nails, 2½-inch

Building a Tip-Up Wall

Partition walls are typically built with 2x4 lumber and have single top and bottom plates. Because it's not structural, the prime requirement of a partition wall is that it support finished wall surfaces. For that same reason, you have the option of spacing studs at 16- or 24-inch centers. Spacing studs at 24 inches on center uses less material, but you must be sure the substrate you use for the finished walls can span the wider distance. In general, ½-inch drywall, the most common material, presents no problem.

1 Mark the Location. Mark the exact location of the wall on the floor. Use a framing square to ensure square corners and a chalkline to ensure straight lines. If you're working alone, wedge one end of the chalkline under a weight or tie it to a nail to hold it as you snap the other end. It's easier to have a helper hold one end, though.

2 Lay Out the Plates. Cut two plates to the length of the wall. Align the plates and measure from one end, marking for studs at 16- or 24-inch intervals on center, whichever you decide to use. Continue to the other end of the plates even if the last stud is less than the required distance from the end. To check your work, measure to a point exactly 48 inches from the end of the plates; done correctly, that point will be centered on one of the stud locations.

3 Cut the Studs. Wall studs measure the height of the wall minus twice the thickness of the framing lumber to account for the thickness of the plates. Subtract an additional ¼ inch from this figure to provide enough clearance to tip the wall into place. Cut all the studs you'll need. To come up with this number, simply count the number of layout marks on the plates. The cuts you make must be square for the wall to fit properly.

4 Build the Frame. Separate the plates by the length of a stud and set them on edge with the layout marks facing inward. Lay all the studs in their approximate positions,

1 Lay out the position of the walls using a framing square and a chalkline. Mark an X on the side of the line that will be covered by the plate.

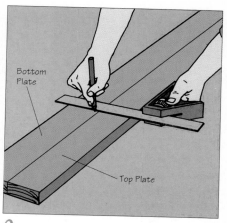

2 Use a combination square to mark the position of each stud on the plates. Mark the plates simultaneously, to be sure the layouts match.

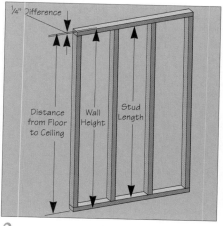

3 Cut the studs 3¼ in. short of the floor-to-ceiling distance; 3 in. for the top and bottom plates and ¼ in. to allow for tip-up clearance.

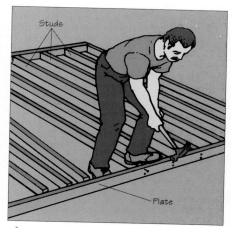

4 Assemble the wall piece by piece, nailing into each stud through the top and bottom plates. Make sure the edges of the studs and plates are flush.

then drive a pair of 16d nails through each plate and into the ends of each stud. Use the marks as guides to align the studs precisely.

5 Form the Corners. To provide a nailing surface for drywall, add an extra stud to each end of the wall that forms an inside corner. One method of making the corner involves nailing spacers between two studs, then butting the end stud of the adjacent wall to this triple-width assembly. Another method is to use a stud to form the inside corner of the wall.

6 Raise the Wall. Keeping an eye on the subfloor layout lines,

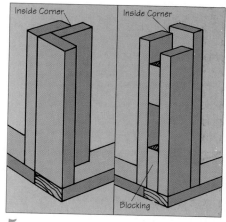

5 Corners must provide a nailing surface for drywall. The left method requires less lumber; the one on the right uses up scraps.

6 *When raising a wall, tip it upright and slide it into position on the layout marks. Nail the bottom plate.*

7 *Use a level to plumb the wall. Check several places on the wall as you nail the top plate. Don't forget to use shims.*

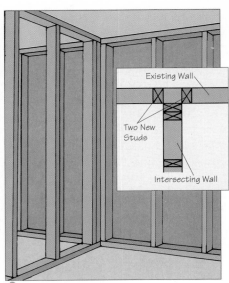

8 *Frame the intersection of two walls so there's a full-length stud on either side of the intersection.*

slide the bottom plate into its approximate position and tip the wall upright. Using a helper to prevent the wall from toppling over, align the bottom plate with the layout lines made in Step 1. Once the wall is in the proper location, nail the bottom plate to the floor. When nailing the plate to a concrete floor, use a single masonry nail set every 18 inches or so. If a wood subfloor is already in place, use the longest common nails possible to nail into the subfloor without hitting the concrete. Nail into sleepers wherever possible. Space the nails in pairs every 24 inches or so.

7 **Plumb the Wall.** Use a 48-inch spirit level to ensure that the wall is plumb, then add shims between the top plate and each intersecting joist (or block) to take up the ¼-inch clearance. Don't over-shim—by doing so, you may bend the top plate out of shape. Just make sure the wall is wedged snugly in place and that it's square, then nail through the top plate and shims into the joists or blocking.

8 **Join Intersecting Walls.** In places where walls intersect,

install additional studs to provide backing for drywall. Add a single stud to the end of the intersecting wall. Add a pair of studs to the other wall. Use 16d nails to secure the intersections.

Building a Wall in Place

It's not always possible to tip a partition wall into place. Pipes or ducts may be in the way, for example, or the wall may be unusually long. Instead, you can build walls in place by slipping each stud between plates that have been nailed to the floor and to the ceiling joists. There are various ways to build a wall in place, but the following method minimizes error:

1 **Install the Top Plate.** First, cut both plates to the length of the wall and mark the top one for the position of the studs at 16 or 24 inches on center. Determine where you want the wall to go and use a 16d nail to attach the top plate to each intersecting joist or to blocking between joists. Be sure that the stud layout faces down so you can see it.

2 **Locate the Bottom Plate.** Hang a plumb bob from the ends of the top plate, transferring the position of the plate to the floor. Align the bottom plate with these layout marks. When it's directly below the top plate, nail the bottom plate to the floor. Use a single masonry nail or anchor every 18 inches or so if you're nailing to a concrete floor, or pairs of common nails every 24 inches if you're nailing to a wood subfloor.

1 *Mark the top plate for the stud spacing of the wall, then nail the plate to the underside of the joists.*

2 Use a plumb bob to locate the position of the bottom plate directly beneath the top plate. When it's in the right place, nail the bottom plate.

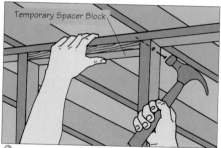

3 Measure between the plates and cut studs to fit. Toenail each stud to the top plate. Use a temporary spacer block to brace the end of a stud.

Install the Studs. Measure between the plates at each stud location and cut studs to fit. The studs will be close to the same length unless the floor or the joists are greatly out of level. Put a stud in position, align it with the layout marks on the top plate, and use a pair of 12d nails to toenail it to the plate. To make toenailing easier, use a spacer block to keep the studs from shifting as you nail them. Cut the block to fit exactly between studs. If the stud spacing is 16 inches on center and the studs are 1 ½ inches thick, the block will be 14 ½ inches long.

Plumb each stud and toenail it to the bottom plate using the same

spacer block. If you're installing the wall directly over a concrete floor, make sure the toenails don't hit the concrete. Use four 8d nails and start the toenailing high enough on the stud so that the end of the nail stops short of the concrete.

Insulating Masonry Walls

The basement—and the garage if it's built into a steep grade—is the coolest part of the house during summer but only moderately cool during the winter. The temperature doesn't vary much because basement walls and tall garage foundation walls are protected from temperature extremes by tons of earth. That coolness feels great on a hot summer day, but to be comfortable in cold weather, most basements and garages require a supplementary heat source. Unless the foundation walls are insulated, much of that supplementary heat is wasted. All walls that face unheated space, such as the wall between a basement or garage recreation room and an adjacent unheated workshop, must be insulated.

There are two basic ways to insulate the foundation walls: One uses fiberglass batt insulation and the other, foam panels. Most likely, it won't be necessary to install insulation higher in insulating value than R-11. Check with local building officials to determine the recommended amount of insulation.

Insulating with Fiberglass

Perhaps the most practical and inexpensive way to insulate a foundation wall is to build a secondary 2x4 wall between it and the living space. This non-load-bearing wall can be insulated with R-11 fiberglass and finished in the same way as other framed walls. Install the wall after you put down the wood subfloor. If there's no subfloor, install the wall directly over the concrete. Use pressure-treated wood for the bottom plate. One advantage to building a secondary wall is that it's easy to run wiring and plumbing lines in it. Also, installing insulation is just as easy as installing it in a standard wall. On the negative side, a secondary wall takes up floor space, which may already be in short supply. You must take care in detailing the framing around windows and doors that are located in basement or garage foundation walls.

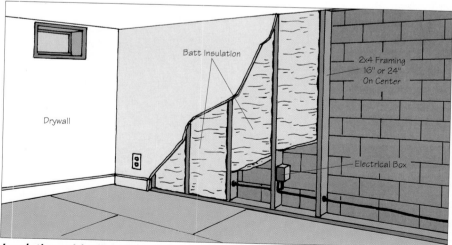

Insulating with Fiberglass. A 2x4 framed wall placed against the foundation walls is a good way to provide insulation. It's also convenient for wiring and plumbing runs.

Before you build the secondary walls, make sure the foundation walls are free from moisture problems. Patch all cracks and use masonry waterproofer to seal the walls. Moisture that collects behind walls eventually leads to problems that are difficult to correct.

Difficulty Level: 🔩🔩 to 🔩🔩🔩

Tools and Materials

- ☐ Goggles
- ☐ Gloves
- ☐ Dust mask
- ☐ Hat
- ☐ Basic carpentry tools
- ☐ Level, 48-inch
- ☐ String
- ☐ Chalkline
- ☐ Framing lumber
- ☐ Wood shims
- ☐ Nails, 10d, 12d, 16d
- ☐ Masonry nails or concrete anchors
- ☐ Fiberglass batt insulation

1 Check the Foundation. A secondary wall can rest directly against the foundation wall, but not all foundation walls are perfectly plumb and straight. To assess the situation, use a 48-inch spirit level to make sure the walls are plumb. Then with an assistant holding one end, stretch a string across the length of the wall and hold it about ¾ inch away from the wall at each end. If the wall touches the string, it's bowed inward; if the gap between string and wall is greater than ¾ inch, the wall is bowed outward.

2 Lay Out the Secondary Wall. You must position the secondary wall so that it's straight and plumb. Doing so might mean that in some places the wall has to stand slightly away from the foundation wall. Locate the innermost line, then measure 3½ inches into the room and snap a chalk line to represent the face of the secondary wall, or the side to which drywall will be attached.

3 Frame the Wall. You frame a secondary wall the same way a basement or garage partition wall is framed, including the option of using 16- or 24-inch on-center spacing for the studs. If you use 24-inch spacing, be sure to buy batt insulation at the appropriate width. Also, remember to allow the ¼-inch clearance that enables the walls to be tipped into place. In other words, cut the studs so they're 3¼ inches less than the distance between the floor and the ceiling joists. If there's a window in the foundation wall, adjust the layout so that there's one stud on either side of it. For the moment, leave out the framing between these two studs.

4 Shim the Top Plate. When you've assembled the wall, tip it into place and align it with the layout

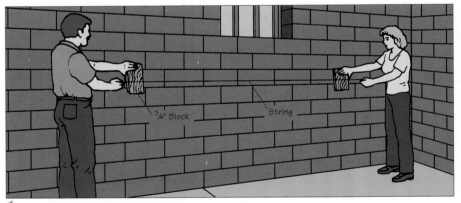

1 Stretch a string across each wall to make sure the wall is not bowed.

2 Snap a chalk line to mark the face of the stud wall. Position the line so that the wall stands clear of high spots.

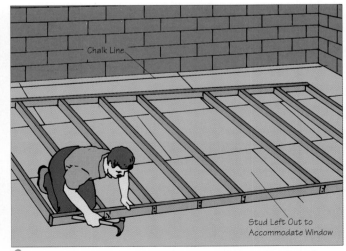

3 Frame the wall as you would a partition wall, but leave out studs as needed to accommodate existing windows.

marks on the floor. Have an assistant hold one end while you stretch a string across the length of the wall. Hold the string about ¾ inch away from the wall at each end. If the wall touches the string, it's bowed inward; if the gap between the string and the wall is greater than ¾ inch, the wall is bowed outward. Use wood shingles to shim between the top plate and the underside of each joist. Make sure the wall is plumb, then nail through the top plate and into the joists using two 16d nails at each location. If the wall is bowed, correct it as much as possible by pulling or pushing on the plate as you nail it.

5 Nail the Bottom Plate. Once you've securely fastened the top plate to the joists, check the position of the bottom plate against the layout lines and double-check the studs for plumb. Then secure the bottom plate to the floor with masonry nails or concrete anchors on a masonry floor or 10d nails on a wood subfloor, just as you would for a partition wall.

6 Frame the Window. Cut a 2x4 sill to fit between the studs on either side of the window. Position the sill ½ inch below the window to allow room for drywall on the sill and nail it through the studs on each side. If there's masonry above the window, you'll also need a header block, set ½ inch above the window. Fill in cripple studs beneath the sill as needed to maintain the stud spacing along the wall. Nail the studs in place through the sill and toenail them into the bottom plate. To let more light into the basement you can angle the drywall away from the window, so reposition the sill accordingly (see "Beveling the Window Sill," page 133).

7 Install the Insulation. The secondary walls must have a vapor barrier on the warm side of the wall.

4 Tip the wall into position (top) and drive shims between the top plate and the joists just until they're snug to lock the wall in place (bottom). Then nail through the plate and shims into the joists.

5 Use masonry nails, screws, or anchors to secure the bottom plate in position according to your layout line.

6 Cut a sill for the window, and nail it in place with 10d nails. Add cripple studs as needed.

The barrier prevents moist air from flowing through the walls and condensing on the cooler surface of the masonry walls. One way to provide the vapor barrier is to insulate with foil- or kraft-faced fiberglass batts. Staple the flanges of the insulation to the studs and eliminate all gaps. You can also install unfaced batts using metal rods called tiger's teeth

7 *Staple the flanges of foil- or kraft-faced fiberglass batts to the studs (left). Don't leave gaps. With unfaced insulation (right), staple 6-mil polyethylene over the face of the studs, overlapping the seams by at least 6 in. Caulk the plastic to abutting surfaces.*

between the studs to keep the insulation in place. After you've installed the insulation, cover the inside face of the wall with 6-mil polyethylene sheeting, lapping the seams for continuous protection. Make sure fiberglass insulation fills all voids in the walls. Don't leave areas of the wall, including the rim and header joists, uninsulated. Insulate behind all plumbing pipes; they must be located on the warm side of the insulation. In areas that experience severely cold winters, doubly protect water-supply tubing with pipe insulation. This material is basically a foam tube with a split down the middle. Simply cut the tube to length, open it up at the split, and place it over pipes. Pipe insulation is available for all sizes of pipe.

CAUTION: When working with fiberglass insulation, protect yourself from the fibers that are inevitably released into the air. Wear a dust mask, eye protection, a long-sleeve shirt, a hat, gloves, and long pants during the installation and whenever insulation is cut or moved.

Installing Rigid Insulation

Rigid insulation has R-values that range from R-4 to R-7 per inch and is made from a variety of plastic materials, including expanded polystyrene, extruded polystyrene, polyurethane, and polyisocyanurate. All of the products come in easy-to-handle sheets, and some are designed specifically for insulating foundation walls. Some products (at least one brand of extruded polystyrene) have rabbeted edges that can be held in place with 1x3 wood cleats. Although extruded polystyrene is highly resistant to moisture, the foundation walls must be dry before they can be insulated.

The following system features extruded polystyrene that's rabbeted on the edges and held in place with wood cleats. The insulation itself is 1½ inches thick and has an insulating value of R-7.5. The sheets are 2x8 feet, so you'll install the cleats on 24-inch centers. You'll also nail the drywall on 24-inch centers instead

of the more typical 16-inch centers. Installing drywall this way works because it's fully supported by cleats and insulation. Check with local building code officials, however, to make sure this insulation method is permitted in your area.

Difficulty Level: ⊤ to ⊤⊤

Tools and Materials

☐ Goggles and gloves
☐ Basic carpentry tools
☐ Rabbeted rigid foam insulation
☐ 1x3s
☐ Electric drill and masonry bits
☐ Masonry screws
☐ Jamb extensions
☐ Caulking gun and caulk
☐ Corner 2x4s

1 **Cut the Sheets.** Measure for the cut and mark it by scoring the insulation lightly with a utility knife. Use the knife to cut through the sheet. It won't cut all the way through, so break the piece off over the edge of a work surface. Use this method to cut each panel so it fits exactly between the floor and ceiling and tightly up against the wall. Cut 1x3 wood cleats to the same height as the insulation.

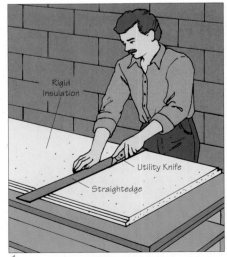

1 *Use a utility knife guided by a metal straightedge to cut rigid insulation. Support the insulation fully on a work-table, then snap it along the cut.*

2 **Place the Sheets.** Start at one corner of the wall. Hold a sheet of insulation against the wall and plumb it. Trim one edge, if necessary, to fit into an out-of-plumb corner. This first sheet determines how plumb adjacent sheets will be, so make sure it's done right.

3 **Install the Cleats.** Hold a second sheet against the first and slip a 1x3 wood cleat into the channel between them. Drill three or four pilot holes through the cleat into the foundation. Pull away the cleat and deepen the holes in the foundation wall as needed. Clear debris from the holes and use masonry screws to secure the cleat. Make sure the heads of the screws are flush with the surface of the cleats. Continue working along the wall in this fashion. Periodically check the insulation for plumb.

4 **Work Around Obstructions.** If you can't move small pipes or other obstructions, work around them by placing cleats on either side. You can fill odd-shaped spaces with expanding spray foam, but it must be a type that's compatible with the insulation. Continue laying out the sheets; make sure the cleats maintain the 24-inch-on-center spacing.

5 **Seal the Edges.** Cut jamb extensions to a size equaling the combined thickness of the insulation and the drywall, and nail them to window and door jambs. Jamb extensions are slender pieces of wood that you nail to the jambs with finishing nails to extend them so they're flush with the finished wall surface. Use a table saw to cut jamb extensions from ¾-inch-thick stock. Once you attach the extensions, use latex caulk to seal small gaps where the insulation meets window or door framing.

6 **Detail the Corners.** You must provide solid support for the edge

2 Start at one corner of the wall. Hold a panel against the wall and plumb it. The panel must fit tightly in the corner.

3 Drill through the cleat into the foundation wall, and use concrete screws to secure the cleat.

4 Use cleats to box-in plumbing pipes and other obstructions. Use expanding foam sealant to fill gaps. Wear gloves and eye protection; don't overfill gaps.

5 Use finishing nails to fasten jamb extensions to the window and door edges. Butt insulation to the extensions and use caulk to seal gaps between them.

of each drywall sheet, particularly at corners. At inside corners, place two cleats edge to edge with a square strip of wood in the corner between them. For outside corners, nailing strips must be the same thickness as the insulation—1½ inches—and at least 3 inches wide so you have some bearing on the basement wall. Two-by-fours make good nailing strips in cases like these. Secure the strips to the corner with screws. Code requires that all areas of rigid insulation that face a living space must be covered, usually by drywall. This is because rigid insulation is combustible; leaving it exposed poses a fire hazard.

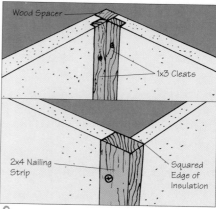

6 At inside corners, use 1x3 cleats and a wood spacer to ensure proper nailing for drywall. At outside corners, nailing strips must be the same thickness as the insulation and at least 3 in. wide.

Enclosing the Garage Door Opening

For a garage conversion to appear as though it were part of an original house design, as opposed to an obvious alteration, both the inside and outside must be finished to look like the rest of the house. The interior must have wall textures and ceiling finishes that match adjacent interior rooms, and the exterior finish must blend with the rest of the house siding, trim, window appointments, and the like. You don't want the garage conversion to look like an afterthought.

As far as the exterior is concerned, there are a number of ways to treat the garage door opening. You can leave the garage door in place and confine the new living area to a space about 8 feet or so behind the door. The extra space will give you a storage area behind the door to stow landscaping tools, bicycles, and the like, and perhaps to install a small workbench. The house will appear to have a garage from the outside, so all would look normal. The fault with this plan, of course, is that you're sacrificing precious living space. Unless you really need a small workbench or storage area in a section of the garage, it may be best to completely convert the garage to living space and either build a carport in front (to make the driveway appear legitimate) or tear the driveway out and landscape that area to blend in with the rest of the yard.

Difficulty Level: 🔨🔨 **to** 🔨🔨🔨

Tools and Materials

☐ Basic carpentry tools
☐ Framing lumber

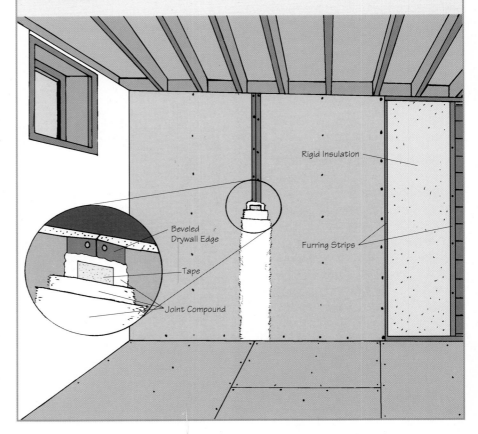

Optional Insulation Method

This method of installing rigid insulation provides the chance to use a greater variety of rigid insulation sheets and allows for a variety of insulation thicknesses. It does not, however, result in an unbroken insulating layer.

First, screw or nail 1x2 wood nailers, called furring strips, to the walls. Then fit sheets of ¾-inch-thick rigid insulation between them and staple a plastic vapor barrier over the assembly. You'll need thicker furring strips if you want to install thicker rigid insulation sheets. A dab of compatible adhesive caulking holds the sheets in place until the assembly is covered with drywall. Be sure that the foundation walls are free from moisture problems before you install the insulation. Drywall or ½-inch-thick (or thicker) wood paneling can be nailed directly to the furring strips.

Rigid Insulation

Beveled Drywall Edge

Furring Strips

Tape

Joint Compound

☐ Nails, 10d, 12d, 16d
☐ Plywood sheathing
☐ Windows or doors
☐ Building paper
☐ Staple gun
☐ Siding to match existing siding
☐ Galvanized 6d or 7d nails

1 Frame up the Garage Door Opening. A concrete stem wall must be installed between the door jambs of the garage door opening to match the existing foundation stem wall, as described in "Closing the Foundation Wall," page 75. The new stud wall will be put in on top of it. There's no need for a header in the new wall, as one already exists to support the structure over the door opening. All you need to do is build a 2x4 wall inside the door opening. Lay out

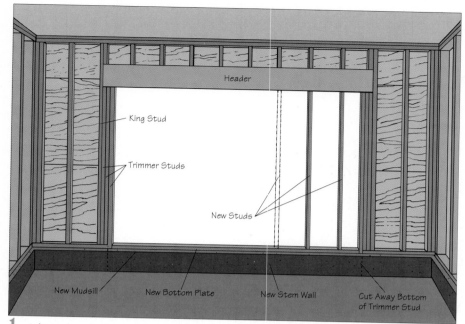

1 A large header is already in place to secure the garage-door opening, so cut the inside trimmer studs where they lap the stem wall, and pour a wall to bridge the opening. Attach a mudsill and bottom plate, then nail the studs in place.

2 Install windows and doors so they butt the top of the original garage door header. Make sure the openings are slightly larger than the overall dimensions of the windows and doors so you can plumb and level the units. After you frame the window, sheathe the garage-door opening.

the wall just as you would a partition wall, except that you'll build the wall in place. You'll need a mudsill and bottom plate but no top plate. Put the studs in position, and toenail them to the bottom plate and the header with 12d nails.

2 **Prepare the Window and Door Openings.** More than likely, you'll want to install at least

one window inside the new frame. Because a header already exists in that space, you won't need to add another. The top of the window should be placed just below the header. The rest of the windows in your house are positioned with their top edges just under the header, so this one should match. Remember, all the finish work in a garage conversion should be done to match the rest of the house. As for any other window, frame the opening according to the rough-opening dimensions that accompany the window, measuring from the existing header to the bottom sill.

Once you've installed the wall, apply sheathing to the outside. Be sure the new sheathing is the same thickness as the existing sheathing so the new exterior siding can be installed flush with the existing house siding. You can measure the sheathing once you've removed the garage door jamb material.

3 **Install the Window.** Inspect the siding around the garage door opening to determine how much existing siding must be removed so that new siding can be installed to make the application look uniform. If your house is sided with vertical tongue-and-groove boards, you'll have to remove the short pieces covering the header area and one or two boards on each side so that the entire wall can be sided to match the siding on both sides of the opening. The same with horizontal siding. Note that horizontal siding boards are staggered along long runs. Butt joints are not aligned. You may need to remove a few boards to install new boards that span the opening in a uniform fashion.

With old siding boards removed, cover the sheathed new wall, including the window opening, with 15-pound felt or air-infiltration-retarding building paper, sometimes known by tradenames such as Typar or Tyvek. Secure the felt or building paper in place with a staple gun or large-head roofing nails. Start at the bottom and overlap seams by at least 3 inches. Use a razor knife to cut an X in the paper at the window opening. Cut from one upper corner down to the opposite side's lower corner. Fold the paper around the window opening and staple it to the inside of the wall to wrap the sides, top, and bottom of the rough opening. Install the window with a helper or two according to the directions provided with window packaging (see "Installing the New Window," page 126).

4 **Install the Siding.** The siding you use to cover the garage door wall insert must match the house's existing siding. If need be, take a section of old siding to the lumberyard to match it exactly. For vertical siding, new boards must be the same thickness and width as the existing ones. For horizontal siding, the clapboards must also match in thickness and width, as well as the bevel angle. Nail tongue-and-groove vertical siding with 6d or 7d galvanized casing nails through the tongue on each board so nails are hidden (except for the last boards, which are face-nailed). Countersink and fill all face nails with caulk before painting. Horizontal "lap," or clapboard, siding is face-nailed along the top or last course with 6d or 7d galvanized casing nails. Countersink and caulk all nails and caulk all butt joints and window perimeters before painting.

There are many styles and types of house siding, so consult with a siding expert at the lumberyard for specific installation instructions. In some cases, especially for aluminum and vinyl siding, you may have to hire a professional siding company to complete the job.

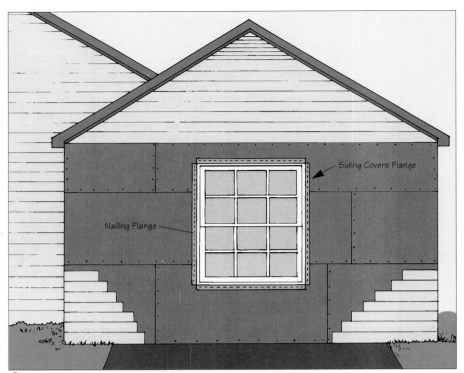

3 *Cover the sheathing with 15-pound felt or building wrap. Tip the windows into the openings. Most windows are nailed through the sheathing and into the framing of a house through a perforated nailing flange surrounding the window.*

4 *Start clapboard siding at the bottom and work your way up, using a chalkline to make sure each course is level. Nail each course at the top of the board; succeeding courses cover the nails. Slip the top of the last course under the existing siding and face-nail it. Caulk the butt joints and window perimeter.*

Doors, Windows, and Skylights

7

Doors and windows play a dual role: They welcome the outside world into your home and protect the inside environment from the elements at the same time. Doors and windows can also add to the architectural style of a home and complement the decor of a room. Of utmost importance with an attic, basement, or garage conversion, though, is uniformity—doors throughout the house should match in style, trim, and hardware. Plan either to match your new doors with those throughout the rest of the house or to replace all existing doors with a new style for an overall home interior face lift.

Door Styles

Along with new interior doors, your attic, basement, or garage conversion may call for a new exterior door. This would include an exterior entrance door opening onto a stair landing if your attic is being converted into a private apartment; a new exterior basement door for both inside and outside appearance upgrades; and an exterior entrance door to a garage conversion for convenient access to a side yard or patio. This section covers both exterior and interior doors.

As you'd see by paging through any door manufacturer's catalog, doors are offered in dozens of shapes, sizes, colors, and materials. Most are made of wood, but you'll also find many doors made of metal, particularly those used as exterior doors. Exterior metal doors often feature a core of rigid foam insulation surrounded by a metal skin. The metal may be embossed or stamped to give it the look of a wood door. Most wood doors are built in one of two ways: As individual panels set in a frame (called a panel door), or as a single plywood sheet, or facing, secured to each side of a wood framework (called a flush door).

Panel Doors. Panel doors offer the widest variety of choices. They can be constructed with as few as three to as many as ten or more panels in all sorts of shapes and size combinations. In some interior doors, especially those for closets, panels may be substituted with louvers, and in some entry doors, the bottom panels may be wood while the top panels are glass.

Flush Doors. Flush doors come in a more limited range of variations and are generally less expensive than panel doors because of their straightforward construction. You can enhance the simple lines of a flush door, however, by applying wood molding to its surface to give it a more traditional look. The doors consist of a surface facing, sometimes called a skin, that covers either a solid or hollow core. Under the facing, the core can be hardwood, particleboard, cardboard honeycomb, or even foam. The rails and stiles are concealed by the facing. The facing is usually made of thin plywood but can also be vinyl- or metal-covered wood, aluminum, or even steel.

French Doors. These traditional doors are framed glass panels with either true divided lights or pop-in dividers. Usually both doors open. Manufacturers also offer units that look like traditional hinged doors but operate as sliders.

Sliding Doors. Patio, or sliding glass, doors consist of a large panel of glass in a wood, aluminum, or vinyl frame. The exterior of a wood frame

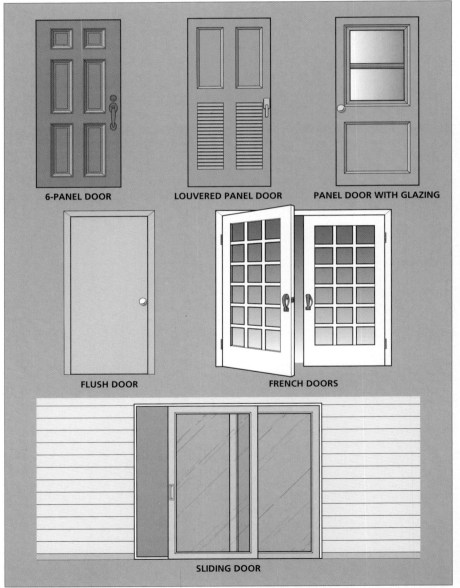

6-PANEL DOOR   **LOUVERED PANEL DOOR**   **PANEL DOOR WITH GLAZING**

FLUSH DOOR   **FRENCH DOORS**

SLIDING DOOR

Door Styles. Doors vary in appearance and construction, as well as in the way they open and close. Their designs vary to suit different functions and architectural styles.

Framing Details

All residential doors hang in a wood frame made of these elements:

■ The door frame, which is nailed within a "rough opening" in the wall formed by wall studs and a header. The framing of interior walls does not always require a header. The rough opening is always large enough so that the door frame can be slipped into place and adjusted vertically or horizontally. Pairs of wood shims are slipped between the door frame and the studs to adjust the door frame. The shims are cut flush with the door frame later.

■ The head jamb, which is at the top, flanked by side jambs, one on the lock side and one on the hinge side.

■ The sill, or threshold, which is often eliminated on interior doors.

■ Stops, or narrow strips of wood nailed to the head and side jambs, which prevent the door from swinging too far when it closes.

■ The strike plate, or the metal strip mortised into the side jamb on the lock side of the door, which accepts the latch.

■ Wood casings at the top and sides, which cover the framework and any gaps, and add the finishing touch to the installation.

■ Weatherstripping, which ideally incorporates interlocking metal strips and should be included all around the frame of an exterior door.

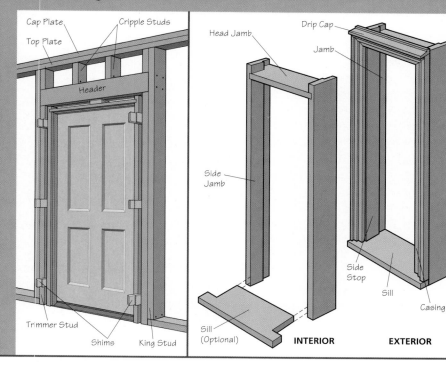

may be clad with aluminum or vinyl. Usually, one side of the door is stationary while the other slides. Because these doors are exposed to the weather, the large expanses of glass should be double-glazed (two layers of glass separated by an air space) to improve the insulation value of the door.

Bulkhead Doors. It's possible to have a door that leads to the outside of the house, even in a basement that's completely below grade. Such a door can provide an emergency exit, although it does not qualify as a bedroom egress under building codes. A bulkhead door also provides convenience. It's easier to get furniture into the basement if the pieces don't have to be lugged through the house. If the basement is going to be used as a shop, a

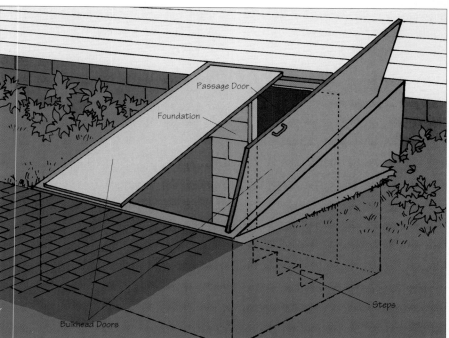

Bulkhead Doors. *The steel panels of the bulkhead door protect the stairwell from weather. An insulated steel passage door keeps heat in the basement.*

door is essential for getting plywood, lumber, and large tools inside.

Adding a door for a basement is a big job: It involves cutting a large hole in the foundation and pouring concrete retaining walls to hold back the earth—a job best left to the professionals. An exterior passage door is framed into the new opening in the foundation. A retaining wall supports concrete or steel stairs, which lead up to the bulkhead door. A bulkhead door typically is made of steel and has two outwardly swinging panels that can be opened from the inside. It caps the stairwell and keeps out rain, snow, and debris. The passage door at the bottom of the stairway must be an exterior-grade insulated metal door with a metal frame. The insulated metal won't rot easily and keeps out the cold.

Building a Frame and Hanging a Door

Making a frame and hanging a door within it takes more time than installing a prehung door but usually costs less. In addition, it allows you to fit the frame to walls of nonstandard thickness.

Difficulty Level: ͳ ͳ ͳ

Tools and Materials

☐ Basic carpentry tools
☐ One-by jamb stock and door stops
☐ Wood shims
☐ Casing nails, 8d
☐ Door
☐ Jack plane
☐ Hinges
☐ Wood chisel, 1¼-inch

1 **Assemble the Jambs.** Purchase nominal one-by jamb stock at the lumberyard or home center. The most commonly available jamb stock is 4½ inches wide to equal the thickness of a 2x4 wall with ½-inch

drywall on each side. You can also buy adjustable split jambs for walls of nonstandard thickness. The side jambs will come with dadoes to accept the head jamb. Allowing for the depth of the dadoes, cut the head jamb to length so that the opening will be ³⁄₁₆ inch wider than the door, and cut the side jambs so that the distance from the top jamb to the floor equals the length of the door plus 1 inch. These dimensions will give the door the recommended ⅛-inch clearance at the latch side, ¹⁄₁₆-inch clearance at the hinge side, ¹⁄₁₆-inch space at the top, and ¹⁵⁄₁₆-inch clearance at the bottom. The bottom spacing allows for ⅛ to ³⁄₁₆ clearance above a ¾-inch finished floor. Nail the side jambs to the head jamb with three 8d nails on each side, and set the frame into the rough opening.

2 **Shim the Jambs.** Level the head jamb as needed by shimming up the bottoms of the side jambs. Plumb one of the side jambs with a 48-inch level, using a pair of opposing shims to adjust the jamb in or out as needed. If you don't have a 48-inch level, use a standard level and hold it against a long, straight board. Nail through the jamb and shims into the trimmer stud with one 8d casing nail. Use three or four pairs of shims: Near the head, in the middle, and near the bottom of the door frame.

Now plumb the other jamb, making sure it's the same distance from the first jamb along its entire length. To keep from having to measure this distance repeatedly, cut a scrap piece of wood to the exact distance required between the jambs and use this "spreader stick" to gauge the opening for placing the shims.

3 **Fit the Door.** If you must trim the top or bottom of a hollow-core door, never remove more than ¾ inch; more will weaken the structure of the door. Use a jack plane to bevel the knob side

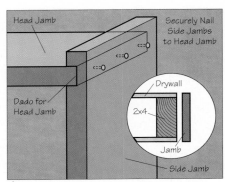

1 Cut head and side jambs, and nail them together. The jambs should be exactly as deep as the wall is thick.

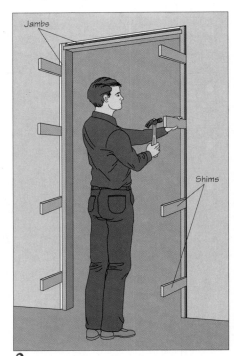

2 Use shims to plumb the side jambs, then nail through the jambs and the shims into the studs.

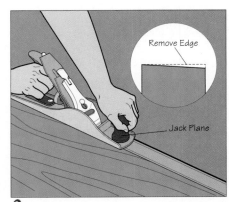

3 After measuring the door's width and height for clearance, bevel the knob side with a jack plane.

(outside) of the new door about 3 ½ degrees; this will help it to close easier.

4 Install the Hinges. The size and number of hinges will vary, depending on the type and thickness of the door. Hollow-core doors, for example, are light and don't require as heavy a hinge as solid doors. Most doors have at least three hinges, though hollow-core interior doors often only have two. Install the top hinge 7 inches from the top of the door and the lowest hinge about 11 inches from the bottom. Center the other hinges. Use a 1 ¼-inch wood chisel to mortise hinges flush with the surface of the door edge, then screw each hinge to the door.

5 Install the Door Stops. Temporarily tack the door stops to the jambs. They should be located so that the hinge side of the door is flush with the jambs.

6 Cut the Jamb Mortises. Now set the door into the jamb, using shims to position it squarely within the opening. Mark the top and bottom of all hinge leaves where they meet the door jamb. Remove the door and use a chisel to mortise the hinge locations into the jamb, just as you did on the door. Many installers prefer to cut hinge mortises before installing the door frame in the rough opening.

7 Complete the Door. Remove the hinge pins, screw the loose hinge leaves to the door jamb, then place the door back into the opening and mate the hinge leaves by inserting the pins. Make sure the door swings freely. If it doesn't, adjust the hinges or door stops, or trim the door. Complete the installation with molding and hardware. When hanging an exterior door, add weather stripping. Paint or finish any door, particularly one leading outdoors, as soon as possible.

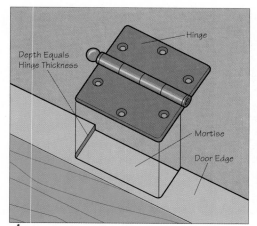

4 Determine the placement of the hinges, then chisel the mortises into the surface of the door's edge.

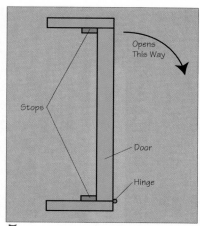

5 Temporarily tack the door stops to the jambs and set the door in place, adjusting it with shims.

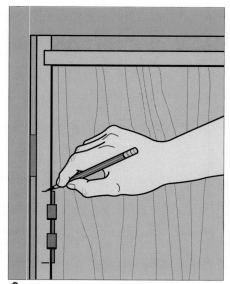

6 Mark the hinge where it meets the side jamb. Loose pin hinges should be installed with the pin at the top.

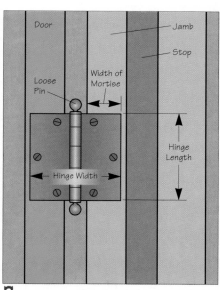

7 To ensure that the door is hung properly, check these details around the hinge. Make adjustments as needed.

Installing a Prehung Door in a Wall

Difficulty Level:

Tools and Materials

☐ Basic carpentry tools
☐ Keyhole saw
☐ Framing lumber and shims
☐ Common nails, 16d
☐ Wood chisels
☐ Casing nails, 8d
☐ Drywall and nails
☐ Prehung door unit

1 Mark and Cut the Door Opening. Use a level and a pencil to mark out the dimensions of the rough opening on one side of the wall. The width of the rough opening is the width of the door plus 2 ½ inches; the height of the rough opening is the height of the door plus 3 inches. If you'll be installing the door in a bearing wall, erect a support structure before removing any framing (see "Build the Support," page 129).

Remove the baseboard and use a keyhole saw to cut through the

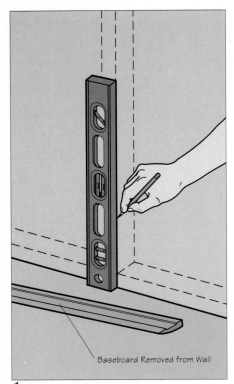

Baseboard Removed from Wall

1 *Using a level and pencil, mark the rough opening, then remove the drywall or plaster with a keyhole saw.*

drywall or plaster along the vertical layout lines, extending the cuts up to the ceiling. Be careful not to cut through any electrical wires or plumbing pipes. Enlarge the opening if necessary to include the nearest existing studs outside the rough opening. These will be the king studs. Remove the wall covering to expose the interior of the wall. Removing plaster or drywall is dusty business, so cover adjacent floors and move furniture well away. Refer to the instructions that come with the door to determine the proper dimension of header lumber to use. Cut off the studs within the rough opening that will serve as cripple studs between the ceiling and the top of the header. Leave the cripples long enough so that the header will be at the proper rough opening height. Remove the bottom portion of the studs from the opening by twisting them from the bottom plate.

2 Make and Install the Header. Assemble the header from two pieces of two-by lumber sandwiching ½-inch plywood cut to the same height and width. Nail the assembly together with 16d nails. The resulting header will be 3½ inches thick, the same thickness as the 2x4 framing lumber. Toenail the header in place, using scrap 2x4 stock to brace it in place while you nail it.

3 Frame and Complete the Opening. Cut a 2x4 to fit between the bottom plate and the header. Toenail this jack stud on one side of the rough opening to support the header. Nail two additional 2x4s together and insert them between the header and bottom plate at the rough opening width. Plumb, then nail them in place.

Use a keyhole saw to cut through the drywall on the other side of the wall, using the framing you just installed as a guide. Cut, fit, and nail drywall in place around the rough opening. With a handsaw and wood chisel, remove the section of the bottom plate inside the rough opening. The framing is now ready to receive the prehung door unit. Nail the unit in place using wood shims as spacers and checking for level and plumb as you work.

Installing Bypass Doors

These lightweight wooden interior doors hang from rollers that slide in a track attached to the head jamb. Bypass doors are most commonly used for closets.

Difficulty Level:

Tools and materials

☐ Door track and hardware
☐ Electric drill and wood bits
☐ Screwdrivers (flat-bladed and Phillips)
☐ Plumb bob

Header

Braces

2 *Make the header out of two two-bys with ½-in. plywood between; brace it and toenail it into place.*

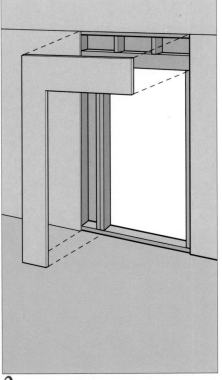

3 *Install studs to create the rough opening width, then nail drywall in place around the rough opening.*

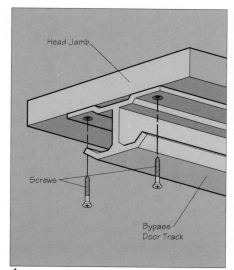

1 *Screw the metal track that guides the bypass doors into the head jamb.*

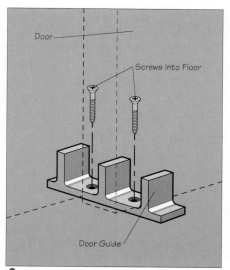

2 *The door guide keeps the doors from swinging back and forth.*

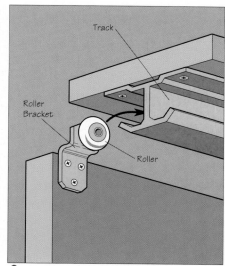

3 *Each door has rollers that slide in the track. Hook the rollers into place.*

1 Install the Guide Track. Once the door frame is in place, which is like any other door opening, screw the guide track to the underside of the head jamb as per the manufacturer's instructions.

2 Install the Door Guide. Install the door guide on the floor, midway between both side jambs and directly below the guide track. Use a plumb bob to find the proper location on the floor.

3 Install the Doors. Lift one door into place and hook its rollers into the back portion of the track. Lower the door into place in the guide track and slide it to one side of the opening. Repeat the procedure with the other door.

Installing Bifold Doors

A bifold door is commonly used on closets, particularly in bedrooms, because it folds neatly out of the way. Each door consists of two panels connected with leaf hinges. Mounting hardware at the top of each door fits into a guide track screwed to the head jamb, while hardware at the bottom fits into a

small pivot bracket that's screwed to the floor.

Difficulty Level:

Tools and Materials

☐ Doors, track, and hardware
☐ Hacksaw
☐ Electric drill and wood bits
☐ Screwdrivers (flat-bladed and Phillips)
☐ Measuring tape

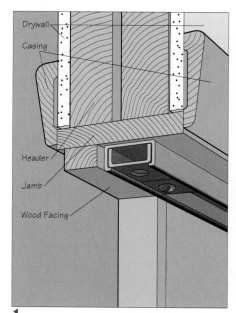

1 *You can conceal the guide track with wood. The wood facing strip doesn't come with the door.*

1 Install the Guide Track. Screw the guide track to the head jamb following manufacturer's instructions. To shorten the track to fit the opening, cut it with a hacksaw.

2 Install the Pivot Bracket. Screw the pivot bracket to one of the side jambs; sometimes there will also be a provision for running a screw into the floor.

2 *The adjustment nut allows you to adjust the door horizontally and vertically.*

3 Install the Door. Install the guiding stud and the top and bottom pivot studs on the doors; holes for these are usually predrilled by the manufacturer. Set the door's top guide and pivot studs into place, then swing the door so the bottom pivot stud aligns with the pivot

3 *Set the upper studs into the guide track, swing the door into place, and drop it onto the pivot bracket.*

bracket and drop the door into place. Adjust the height of the door by turning the pivot in the bottom bracket clockwise to raise the door or counterclockwise to lower it.

Door Casing

Casing can be simple or complex, depending on the wood, joinery, and detailing. Door molding is easy to recognize, and it can be manipulated in dozens of ways.

▶ Head casing is the horizontal member that spans the top of the door frame.

▶ Side casing consists of the molding on the sides of the door. Like the head casing, side casing is nailed into the edge of the door frame.

▶ Corner blocks are decorative blocks that can be used to make the transition between horizontal members and vertical members.

▶ Plinth blocks are used at the bottom of the side casing where baseboard meets the door trim.

Installing Mitered Door Casing

There are several ways to join head casing to side casing around a door. The most common method is to use a miter joint, which is simply two 45-degree angles joined to make a 90-degree angle. You can use other joints, however, depending on your skill and the architectural style of your house.

Difficulty Level: 🔨🔨

Tools and Materials

- ☐ Combination square
- ☐ Pencil
- ☐ Measuring tape or folding ruler
- ☐ Miter box and backsaw (or power miter saw)
- ☐ Lightweight trim hammer
- ☐ Casing nails, 3d and 4d
- ☐ Nail set
- ☐ Wood putty

1 Mark the Reveal. The inside edge of the casing should be offset from the inside edge of the jambs by approximately $3/16$ inch. The small edge caused by offsetting

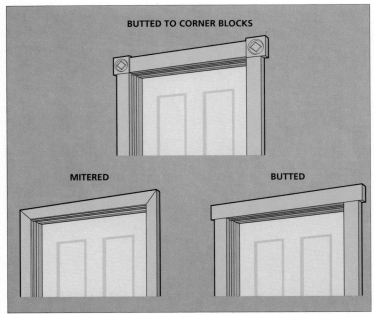

Door Casing. *The molding around a door may be joined with miter joints or butt joints. Butted corners may use decorative corner blocks.*

BUTTED TO CORNER BLOCKS

MITERED

BUTTED

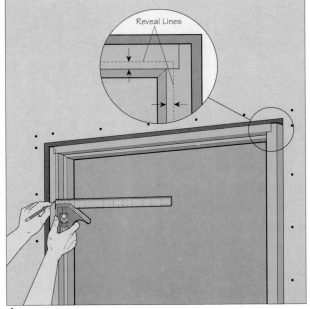

Reveal Lines

1 *Use a combination square to mark the $3/16$-in. reveal on the jambs.*

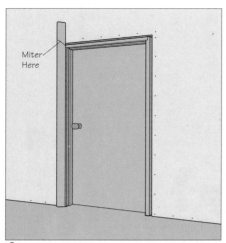

2 *Miter where the side casing reveal intersects the head casing reveal.*

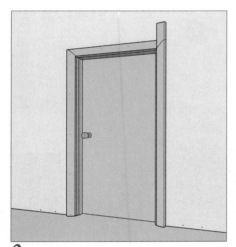

3 *Cut the head casing to size, then mark and cut the second side casing.*

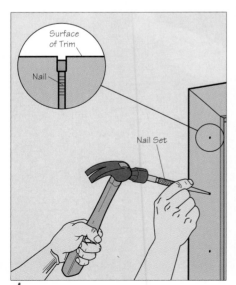

4 *Set all nails below the surface of the casing with a nail set and hammer.*

the two is called a reveal. Set the combination square for ³⁄₁₆ inch and use it to guide your pencil around the jamb, leaving a line ³⁄₁₆ inch from the edge.

2 Make the First Miter. Cut a length of casing square at one end. Then place the casing against the reveal line with the square cut against the floor. Mark the casing at the point where the vertical and horizontal reveal lines intersect, and cut a 45-degree angle at this point with a miter box and backsaw or power miter saw.

3 Make the Next Miters. Tack the first piece of casing to the jamb with 3d or 4d casing nails. Don't drive the nails home. Now cut a 45-degree angle on another piece of casing for the head casing, fit it against the side casing, mark it for the opposite 45-degree angle, then cut and tack it in place. After

attaching the head casing, mark, cut, and install the final length of side casing.

4 Set the Nails. Finish nailing the casing in place with nails spaced about every 12 inches, set all the nails just below the surface of the wood, then fill the holes with wood putty and sand them smooth when dry. Use a nail set and a lightweight hammer to set the nails. In a pinch, carefully use a 10d nail as a makeshift nail set.

Door Hardware

Every door needs hardware, which is available in many styles, from sleek and contemporary to detailed traditional designs, and in various materials from burnished aluminum to solid brass. Although the most

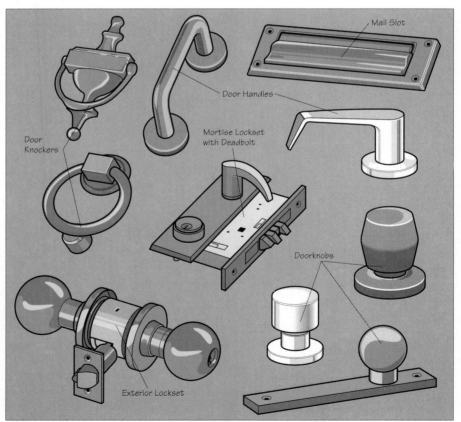

Door Hardware. *Door hardware is available in a multitude of models and styles. Try to find hardware for new doors to match that in the rest of the house.*

important requirement for hardware is that it be functional, it must match the rest of the door hardware in your house as well. When purchasing hardware, look for heavy-gauge metal, fine machining without sharp or rough edges, and a plated finish to withstand heavy use.

Match the hardware to the door. A solid, heavy entry door, for instance, must be hung with heavy-duty hinges. Think about security when selecting hardware like locksets and dead-bolt assemblies for exterior doors.

Locksets

The term "lockset" refers collectively to the complete door-latch system: latch-bolt assembly, trim, and handles, knobs, or levers. A latch bolt is a spring-loaded mechanism that holds a door closed and may or may not have a lock incorporated in it. A deadbolt, on the other hand, isn't spring loaded and can be operated only with a key or a thumb-turn. The following steps explain the basic procedure for installing a lockset, although you should always refer to the instructions that come with most sets.

Installing a Lockset

Difficulty Level:

Tools and Materials

☐ Basic carpentry tools
☐ Lockset
☐ Awl
☐ Electric drill with assorted bits and hole saws
☐ Wood chisel, ¾-inch

1 Locate the Holes. Using an awl and the template included with the instructions, mark positions for the knob assembly holes. The knob should be 36 to 38 inches from the floor. Its hole should be 2 ⅜ inches or 2 ¾ inches from the

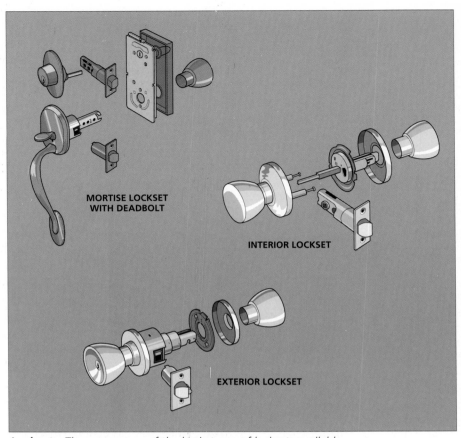

MORTISE LOCKSET WITH DEADBOLT

INTERIOR LOCKSET

EXTERIOR LOCKSET

Locksets. These are some of the basic types of locksets available.

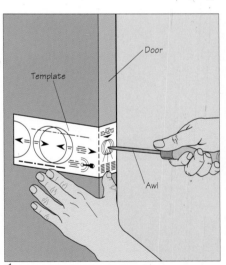

1 *Use a template to mark the hole location. To measure accurately, fold the template around the door.*

edge of the door, depending on the lock.

2 Drill the Holes. Bore a hole (the size specified for the lock tube) into the face of the door.

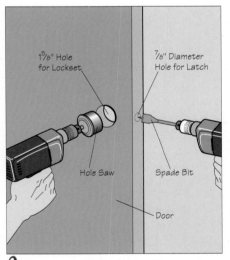

2 *Bore holes in the door with a power drill. Use a hole saw for the lockset hole and a spade bit for the latch hole.*

Drill first from one side, then from the other to avoid splintering the wood. Next, drill a hole into the edge of the door for the latch and assembly.

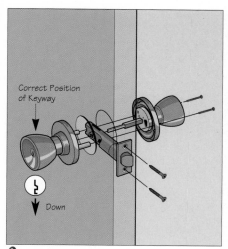

3 *When installing a keyed lockset, align the keyway as shown.*

4 *Mark the plate's position on the door to help align the plate on the jamb.*

5 *Remove the plate and close the door. Find the location of the strike.*

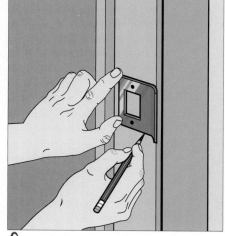

6 *Mark the position of the strike by tracing the strike plate on the jamb.*

3 Install the Lockset. Insert the cylinder assembly and latch into the door. Mortise the latch plate into the door using a ¾-inch wood chisel.

4 Mark the Door Edge. Place the strike plate over the door latch and mark the plate's position on the edge of the door as reference marks for when you cut the strike plate into the door jamb with the chisel.

5 Center the Latch. With a sharp pencil, pinpoint the spot where the center of the latch hits the door jamb.

6 Mark the Strike Location. Hold the strike plate to the door jamb, centering the hole over the pencil mark made for the latch. Also, make sure the plate is flush with the top and bottom marks you made on the edge of the door. Trace the location of the strike plate and the latch on the door jamb. With a sharp chisel, cut a mortise into the jamb equal to the depth of the strike plate. If you make the cut too deep, use cardboard to raise it so it's flush. To make room for the latch, use a drill or chisel to bore a hole into the center area of the strike plate. Fasten the strike plate to the jamb with screws, checking the alignment again.

Installing and Trimming Windows

There are five common types of windows: Fixed, double-hung, casement, awning, and sliding. Only double-hung and casement windows are appropriate for most attic projects. Awning windows are generally used in basements. You can install any kind of window in a garage conversion, depending on your needs. Windows can be used individually or combined in various ways to achieve dramatic effects.

▶ Fixed, or stationary, windows are the simplest kind of window because they don't open. A fixed window is simply glass installed in a frame that's attached to the house.

▶ Double-hung windows are perhaps the most common. They consist of two framed glass panels called sash that slide vertically and are guided by a metal or wood track. One variation, called a single-hung window, consists of a fixed upper sash and a sliding lower sash. The sash of some double-hung windows can be tilted inward to make it easier to clean the outside. This is particularly handy for attics because otherwise windows are accessible from the outside only by ladder.

▶ Casement windows are hinged at the side and swing outward from the window opening when you turn a small crank. They can be opened almost completely—90 degrees from the closed position—for maximum ventilation. More importantly, casement windows may be used as egress windows because most of the sash area is unobstructed when the window is opened.

▶ Awning windows are similar to casements in that they swing outward, but they're hinged at the top. A useful feature of the window is that it can be left open slightly for ventilation, even during a light rain. One variation, call a hopper window, is hinged at the bottom and may be opened outward or inward.

▶ Sliding windows are like a double-hung window turned on end. The glass panels slide horizontally and are often used where there's need for a window that's wider than it is tall.

Difficulty Level: 🔩🔩 to 🔩🔩🔩

Tools and Materials

☐ Basic carpentry tools
☐ Cat's paw
☐ Wrecking bar
☐ Hacksaw
☐ Common nails, 12d
☐ Chalkline
☐ Goggles
☐ Circular saw
☐ Windows
☐ Shims
☐ Caulking gun and caulk
☐ Finishing nails, 4d, 6d, 10d
☐ Flashing
☐ Insulation
☐ Staple gun and ½-inch staples
☐ Building paper
☐ Roofing nails, 1¾-inch
☐ Framing lumber
☐ Miter box and backsaw (or power miter saw)
☐ Casing
☐ Nail set
☐ Wood putty

Installing Gable-End Windows

With the exception of skylights and dormer windows, attic windows are installed in gable ends. Often, an attic already has a window in one or both gable ends. It may be an operable unit originally intended for ventilation purposes or a fixed unit

Windows. *These are the five most common kinds of windows. Each one lends its own charm to a room.*

FIXED

DOUBLE-HUNG

CASEMENT

AWNING

SLIDING

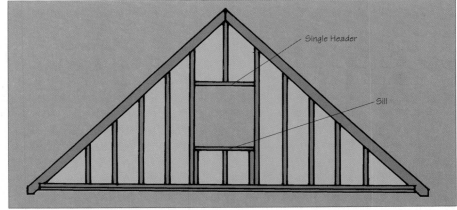

Single Header

Sill

Installing Gable-End Windows. *When the window is in a non-bearing gable wall (and it is not unusually large), the framing can be simplified.*

that simply allows a bit of light to penetrate. In either case, it's most likely a single-glazed unit and not suitable for a living space. It's easier to install windows in gable ends than in other parts of the house. This is because the framing at the gable ends is rarely load-bearing; all the load is carried by the rafters.

If windows already exist in the gable end, the framing around them is probably lighter than that found in the outside walls below. Instead of a heavy structural header, a single or double plate might frame the top of the window. If this is your situation, you have complete flexibility in changing the framing to suit your window plan. Also, there's no need to provide temporary support for the wall while you change framing because the wall isn't load-bearing. The only thing that adds difficulty to working on gable-end windows is the fact that you're working high above the ground.

Structural Ridgebeams. If the attic has a structural ridgebeam, there's a possibility that the gable end is load bearing. In this case, the ridgebeam supports the top end of each rafter, rather than the other way around. The ridgebeam, in turn, is supported

by posts that run down the middle of each gable end. Structural ridge-beams most often are used for cathedral-style ceilings because they eliminate the need for joists to keep the rafters from spreading. One clue to this situation is a ridgebeam that's unusually heavy, such as built up 2x10s or a glue-laminated beam. Also, the rafters may be notched to sit on top of the beam instead of simply being nailed to the beam. If you think you may have a structural ridgebeam, consult a structural engineer, architect, or master framing carpenter before disturbing the gable-end framing.

Framing the Window

All windows fit within an opening in the wall framing called a rough opening. The rough opening is slightly larger than the overall dimensions of the window measured at the jambs. The extra space allows the window to be plumbed and leveled as needed. The exact width and height of the rough opening is specific to the particular window you buy. It's usually best to wait until you have the window in hand before making the opening. The following procedure assumes that

the wall isn't load bearing and that there's no existing window in the gable end; however, the procedure is essentially the same for enlarging a window opening.

1 Remove the Studs in the Way. Start by removing all the studs that cross the area to be occupied by the rough opening. The studs meet the end rafters with angled cuts. In most cases, the studs are toenailed into the rafters. Use a cat's paw to remove the nails. Next, release the studs from the nails coming through the bottom plate. There are two ways to do this: If you have a reciprocating saw, fit it with a metal-cutting blade and cut in places where the studs meet the plate, severing the nails. If you don't have a reciprocating saw, use a handsaw to cut through the studs about 4 inches above the plate. The sheathing and the siding are nailed from the outside into the studs. Use a wrecking bar to pry the studs away from these nails. Save the studs; they can be used for framing the new rough opening. If you used a handsaw to cut the studs, knock the remaining pieces free of the plate nails, then use a hacksaw to cut the protruding nails flush. Cut the nails that protrude through the sheathing, as well.

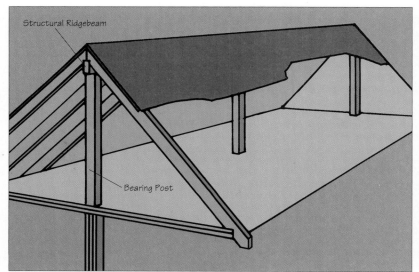

Structural Ridgebeams. *Attics that have a structural ridgebeam also have bearing posts at the gable ends, so adding windows is more complicated.*

1 *Remove all the studs in the area of the new window for non-bearing gable-end walls.*

2 *Install the king studs that form the sides of the rough opening.*

3 *Install the new header, sill, and cripple studs.*

4 *Tack a board to the siding as a guide for cutting the opening.*

2 **Install the New King Studs.** Lay out the width of the new rough opening on the bottom plate. Use a long level or a plumb bob to transfer the width to the bottom of the end rafters. Measure at each mark to get the lengths of the king studs, which define the new rough opening. The studs can probably be cut from studs that have been removed; they already have the proper angled cut on the top ends. Toenail these studs to the bottom plate and rafters. Check for plumb.

3 **Install the Header, Sill, and Cripple Studs.** Cut the header and rough window sill to fit between the designated studs. Attach them by nailing through the studs. If there's not enough room to swing the hammer, toenail the header and sill to the studs. The cripple studs are the short vertical members below the sill and above the header. In a non-structural wall, their only purpose is to provide a nailing surface for dry-wall inside and sheathing outside. Space the cripple studs as needed for such nailing—usually 16 inches on center.

4 **Cut the Opening.** From inside, drill a hole at each corner of the opening. Then go outside and snap chalk lines from hole to hole to delineate the opening. All nails that are crossed by the chalk line must be removed prior to cutting. If the siding is horizontal clapboard, you must provide a level surface upon which the circular saw can ride by tacking a one-by board along the cut line as shown. Set the saw to cut through the siding and sheathing. Wear safety goggles, and cut the opening.

Installing the New Window

You can install windows in any of several ways. Some windows are installed by driving nails through the jambs and into the sides of the rough opening. These windows have a narrow casing called brickmold on the outside. Metal, vinyl, aluminum-clad wood, vinyl-clad wood, and some all-wood windows are secured by nailing through a perforated flange that surrounds the win-

dow into the sheathing and framing below. With this system, you have to cut back the siding (but not the sheathing) to make room for the nailing flange. You also need a window casing, which is used to cover the nailing flanges. In another system, the windows come with a heavier outside casing that gets nailed to the outside of the rough framing. You have to cut back the sheathing as well as the siding.

The following directions pertain to the type of window that gets nailed through the jambs.

1 **Put the Window in Place.** Unpack the window and check it for square by measuring the window jambs corner to corner. The diagonal measurements must match. If there are any braces or reinforcing blocks on the window, leave them in place until the window has been nailed securely to the house. Some manufacturers recommend that the sash be removed before the window is installed to prevent glass breakage, while others recommend leaving the sash in place to stiffen the jambs. If

the manufacturer permits, remove both sash to make the window easier to carry up the ladder. Lift the window into the rough opening and hold it there firmly as an assistant helps from the inside.

2 **Level the Window.** Check the sill for level. Shim beneath each jamb leg as needed. If the window is unusually wide, shim the sill midway between legs as well. Check the sill frequently throughout the installation to ensure that it has not shifted out of position. Tip the

window away from the opening just enough for your assistant to run a bead of exterior-grade caulk behind the brickmold, then press the window against the wall.

3 **Set the Window.** Use a 10d galvanized casing nail to nail through the casing, securing one lower corner of the window to the wall. Casing nails are similar to finishing nails, but their shanks are heavier. Insert flashing over the head casing and slip it beneath the siding. If flashing did not come with

the window, buy it at a building-supply store.

4 **Plumb the Window.** Check the window for plumb; this is particularly important if you removed the sash earlier. Check the window again for square. If necessary, adjust it by slipping shims between the jamb and the framing. When the window is plumb, use another nail to tack it in place.

5 **Nail the Window.** Install both sash, and open and close them

1 Remove the sash from a double-hung window before installation to make it easier to lift.

2 Have an assistant shim the window while you check it for level and plumb from outside.

3 Don't set the nails until the window has been installed. Slip flashing beneath the siding and over the head casing.

Siding

Flashing

Head Casing

4 Plumb and shim the jamb from inside the attic as needed. Check the window for square as well.

5 Install the sash. If they slide smoothly, finish nailing the window into place.

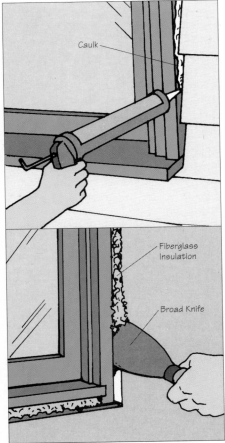

6 Use caulk to seal around the outer edges of the window (top), except at the head flashing. Use fiberglass insulation or expanding foam to seal gaps on the inside (bottom).

a few times. If they work properly, complete the nailing. If the sash bind, however, you may have to reposition the window.

6 **Seal the Window.** Use a high-quality exterior-grade caulk to fill the gap beneath the sill and between the window casing and the siding. From the attic side of the window, use fiberglass or a foam sealant to seal the gap between the jambs and the rough opening. Foam expands as it cures, so spray it gradually to prevent it from pushing the jambs out of position. The window is now ready for you to install the interior casing.

Installing Flanged Windows

At one time, the only windows with a perimeter nailing flange were metal windows. Now, however, there's a flange on windows made of aluminum, vinyl, vinyl-covered and aluminum-covered wood (clad windows), and some wood windows. Rough openings for all flanged windows are typically smaller than they would be for a standard window of the same size. This is because there's no need to shim the sides of the window. That makes installation easy. The following steps detail basic installation. Always refer to the instructions included with your window, however, because details vary according to the manufacturer.

1 **Set the Window in the Rough Opening.** When putting a window in a new wall or an existing wall with no window, consult the manufacturer's literature to determine the dimensions of the rough opening. Once the framing is complete for the new wall, nail the sheathing into place around the rough opening. It may be easiest to cover the rough opening initially, then cut away sheathing from the rough opening afterward. Once window framing

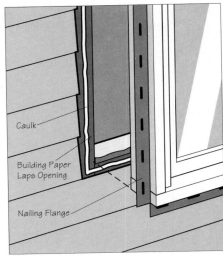

1 As you set the window into place, run a generous bead of caulk between the flange and the sheathing.

2 Be sure the sill is level, and shim if necessary. Begin nailing at one corner.

is complete on the inside of an existing wall, cut away sheathing as explained in Step 4 of "Framing the Window," page 126. Staple or nail 15-pound building paper over the sheathing, and wrap it around the top, bottom, and sides of the rough opening. Caulk the perimeter of the opening, then place the window in the rough opening.

2 **Level the Sill.** Check the sill for level, and shim beneath the window from inside the house as necessary. Begin nailing the window

3 *Drive nails partway in the other corner, checking for plumb and level before you drive them home.*

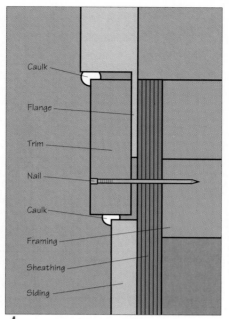

4 *Nail up the trim and caulk it along both edges. Try to place nails to avoid the nailing flange of the window.*

in place on one side, using 1 ¾-inch roofing nails through existing slots in the flange.

3 Nail the Window. Nail through the flange on the other side of the window, checking for plumb. The best technique is to drive the first few nails partway, fully driving them home only after you're sure the window is plumb and level. After several nails are in place, check the operation of the window by opening and closing it several times. Don't nail through the top nailing flange. Rather, start the nail just above the flange to allow the head to catch the top edge of the flange (or use 8d nails and bend them over the flange). This is done in case the header above ever sags. If nails at the top of the window were placed through the top flange and into the header, pressure from a sagging header would be transferred to the window frame, causing the window to bind or the glass to crack.

4 Finish the Window. Install trim or replace the siding around the window as needed to match the

other windows of the house. The casing should fit snugly between the edges of the siding and the side of the window. Caulk in both places. If head flashing was supplied with the window, install it now.

Installing Garage Conversion Windows

Installing windows in the walls of a garage conversion is no different from installing windows in most rooms, with the exception of attics and secondary walls in basements. Framing windows for the new wall insert at the garage door opening is easy, as discussed in Step 2 of "Enclosing the Garage Door Opening," page 111. What's more difficult is installing windows in load-bearing walls away from the garage door opening.

Once you've selected a window location and marked the rough opening layout on the wall, use a keyhole saw to remove any drywall that may be present. Before you begin to remove studs and

start framing, however, you must support the ceiling joists and all of the structural members above the proposed window opening. The support will remain in place until all rough opening framing work is complete, including the installation of a properly sized header.

Build the Support. Erect a support structure made up of a header and two posts using 4x4s or double 2x4s. Set the support no more than 24 inches away from the existing wall in line with the proposed window rough opening and about 24 inches past each side to ensure that all affected ceiling joists above are properly supported during the framing procedure. Cut the posts so they'll fit snug. Support one end of the header with a post while a helper holds the other end. Then set the other post in position. Once the support system has been erected, you can begin framing work for the window's rough opening (see "Framing the Window," page 125).

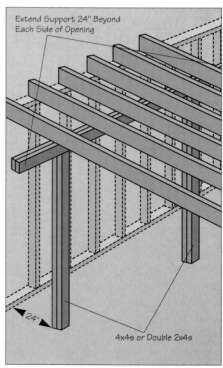

Build the Support. *Before removing a bearing wall, you must support the ceiling joists with a temporary 4x4 structure.*

Install the Header. The one exception to the window-framing process mentioned earlier is the necessity of installing a header for the window. To make the header, sandwich a length of ½-inch plywood between two two-by boards. In a garage with only a roof above, the size of the header will be determined by the span of the opening. For a span of 48 inches, use two 2x4s; for a span of 72 inches, use two 2x6s; for 96 inches use two 2x8s; for 10 feet use 2x10s; and for 12 feet use 2x12s. In a wall with living space above it, consult an engineer for the required size of the header. Set the header between the king studs and on top of the trimmer studs.

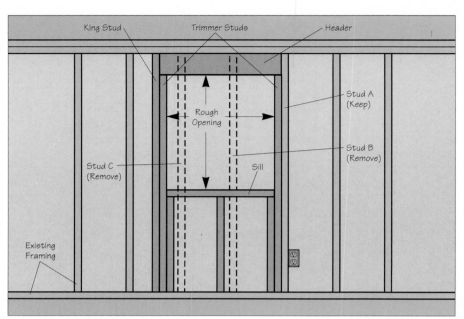

Install the Header. *Support the span of the window opening with a doubled two-by header set between the king studs and supported by trimmer studs.*

Installing a Window Stool

A "stool" is a piece of wood that's placed on top of the sill as part of the trim work. Stools can be made wide to support small plants, photos, and the like. The stool should be installed before the casing and the apron—molding that goes under the stool. Some stools have an angled underside that matches the sill, while others are flat to match flat sills. In either case, installation is the same.

1 Cut the Stool. First, cut the stool to length. Generally the "horn" of the stool extends slightly beyond the casing on both sides. This is primarily an aesthetic decision, however, so make the horn to your own tastes. Mark the center of the stool and make a corresponding mark on the center of the window frame.

2 Mark the Cut Lines. Hold the stool against the window jambs and align the two center marks. To lay out the horns, slide a combination square along the front edge of the stool until the blade rests against one side jamb of the window. Mark the stool as shown. Repeat this on the other side of the stool.

3 Trim the Stool. While holding the stool against the jambs, measure from its inside edge to the

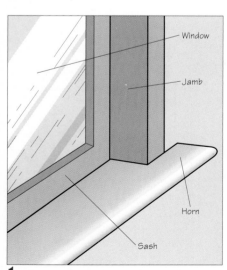

1 *The stool sits against the window sash, with its horns extending beyond the casing by ½ to 1 in.*

2 *Square across each end of the stool and draw a line corresponding to the inside face of the jamb.*

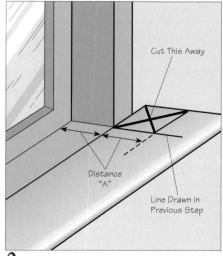

3 *Measure from the stool to the sash ("A"), then mark this distance along the line drawn previously.*

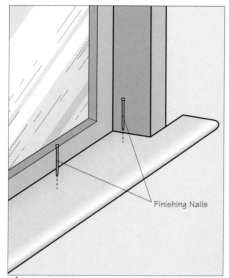

4 Sand the edges of the stool, then set it into place and nail it to the window frame with 6d finishing nails spaced every 10 in. or so apart.

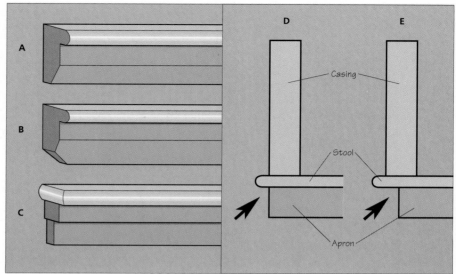

5 With the ends cut square (A), the end grain shows. With the ends cut square and the lower corners cut at 45-degree angles (B), the end grain also shows. With the ends mitered and "returned" (C), no end grain shows. The ends of the apron either line up with the outside edge of the casing (D) or stop at the midpoint of the casing (E).

sash. Transfer this measurement to the marks just made in the previous step, then draw a perpendicular line from each point to the end of the stool. Cut the stool along the layout lines, using a handsaw or saber saw. The stool should now fit into place on the window sill.

4 **Nail the Stool.** Round the ends of the horns to match the face of the stool. Nail the stool into the window framing with 6d finishing nails spaced 10 inches or so apart, and set the nails.

5 **Install the Apron.** The apron is the simplest part of the window trim to install, but even here there's room for creativity. First, decide on how long to make the apron and on how to deal with its cut ends. Both decisions rest largely on personal taste. In terms of length, you may prefer to line up the ends with the outside of the casing or have them fall just short of the casing. The difference between the two amounts to fractions of an inch but may have a significant visual impact. If the rest of the windows in your house are outfitted with stools and aprons,

match the new windows with those. The ends of the apron can be cut square and sanded smooth or finished off with a flourish. Again, it's more of a personal preference and should be based on the existing window trim in the rest of your house.

Installing Casing around a Window

Casing a window is the fussiest part of installing a window—any imperfections in the joints will be readily apparent. Making accurate joints may take some practice, but the process isn't difficult.

1 **Mark the Reveals.** The first step is to mark out reveals on the edges of the jambs. A reveal is a slight offset between the inside face of the jamb and the inside edge of the casing (just like door casing). It's easier to install the casings when you don't have to make them perfectly flush with the inside face of the jamb, and they look better, too. To mark the reveals, adjust the blade of a combination square to the size of the reveal (about $3/16$ inch) and mark the jambs as shown.

2 **Cut the Side Casing.** Each side casing will have a miter at one end and a square cut at the bottom end if the window sill is trimmed with apron and stool. Otherwise, side casings will have both ends mitered so they meet with casings at both the top and bottom. To find out how long the side casing should be for a window with a stool, set the head casing in place, aligned with the

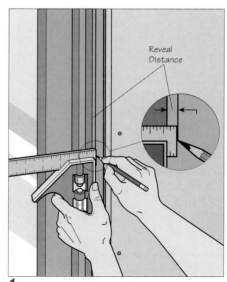

1 Mark the reveals along the edges of the window jambs. The reveal distance should be about $3/16$ in.

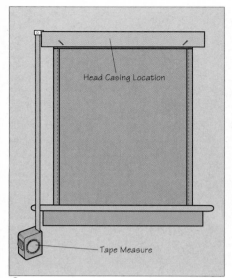

2 Tack the head casing. Measure from its outside tip to the top of the stool for the length of the side casing.

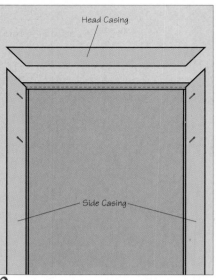

3 Cut the head casing with a 45-degree miter at each end to meet the side casing just mitered.

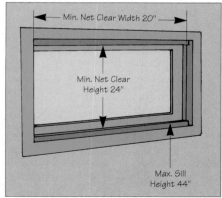

4 Nail the casing with 4d finishing nails into the jamb, and 6d nails through the drywall into the framing.

reveal mark, and measure from the top of the head casing to the top of the stool. Cut one piece of side casing to size, with the bottom square. Place the piece along the side jamb reveal line and mark it where it intersects the head casing reveal line. Make the miter cut at this point. Repeat for the other piece of side casing. Tack the side casing in place, lining it up with the reveal marks. Use 4d finishing nails for nailing into the jambs and 6d nails for nailing into the framing. Leave the nailheads sticking out for now to reposition the casing if necessary.

3 Cut the Head Casing. Measure between the side casings and cut the head casing to this dimension with both ends mitered to meet the side casing.

4 Nail the Casing. Align the side casing with the head casing and trim it as necessary for a perfect fit. Cut and test the other side casing. If all the joints look tight, you can nail everything in place. Nail the casing to the jambs and to the studs behind the drywall. Set the nails with a nail set and fill the holes with wood putty.

Installing Basement Windows

When it comes to basement windows, the key is to make the most of what you have. First, repair or replace damaged windows. Then, trim the windows to fit the decor of the new room, preferably to match the windows in the rest of the house.

Unless yours is a daylight basement (a basement with at least one wall exposed to the grade), it probably does not have much by way of windows. For a recreation room or a home office, this isn't a problem. Simply wire the room so that it has plenty of artificial light. For rooms used as bedrooms, however, building codes come into play.

Building Codes and Basement Windows. Whether or not a window has to be added to a basement isn't always just a matter of personal preference. Building codes require that all bedrooms, including those in a basement, have a means of emergency egress.

If there's a door that leads directly outside (and not to a bulkhead door), it can be considered an

emergency exit. If there's no door, however, each bedroom must have at least one egress window. The requirements for such a window specify its height (24 inches, minimum), its width (20 inches, minimum), its "net clear opening" (5 square feet, minimum), and the maximum distance between the sill and floor (44 inches). The net clear opening is essentially the amount of space available for a person to climb through once the window is fully open. It's measured between obstructions, such as window stops, that restrict passage.

Building Codes and Basement Windows. An egress window is used as an emergency exit. The net clear opening can't be less than 5 sq. ft.

Replacing a Wood Window

One problem with wood-framed basement windows is that they're susceptible to rot and insect damage. In some cases the affected wood can be simply cut away and repaired with epoxy wood filler. Extensive damage calls for replacement of the entire window. Measure the size of the rough opening and check local window sources to see whether replacement windows are readily available. If not, a custom-built window has to be ordered. This may take some time, so don't remove the old window until the new one is in hand. Replacement methods vary depending on how the original window was installed; so pay attention as you remove the old one.

Window Details. With most basement windows, the frame is flush with the inside surface of the foundation wall. If the inside of the foundation is to be insulated, however, the window has to be "boxed-out" so it matches the combined thickness of insulation and finished wall surfaces. Given the variety of window sizes, frame types, and locations, there's no one best way to box a window. If the window provides egress, check with local building officials before proceeding—boxing out a window sometimes affects its accessibility. The following are some options to consider.

Boxing a Window

For wood-framed walls that are built to insulate the foundation (secondary walls), there are several ways to finish the area around the foundation windows. The simplest way is to treat the window as if it were in a standard frame wall. A 2x4 sill nailed between studs forms the rough opening while the ceiling butts into the top of the window. The new jambs and sill may be

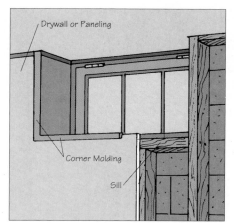

Boxing a Window. *In places where a secondary wall meets a window, you can detail the framing to support finished wall surfaces.*

finished with paneling or drywall. Remember to consider their thickness when you install the framing.

Beveling the Windowsill

An alternative to boxing out a recess is to bevel the windowsill. Beveling the sill takes planning and some carpentry skills but results in a brighter basement because it allows sunlight to spill into the room.

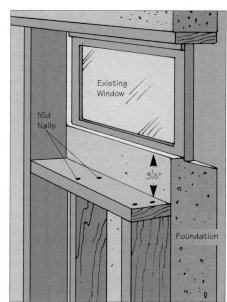

1 To provide support for a beveled sill panel, frame the wall below the window so that its top plate is lower than it would otherwise be.

Difficulty Level:

Tools and Materials

- ☐ Basic carpentry tools
- ☐ Framing lumber
- ☐ 1x4s
- ☐ Masonry screws or anchors
- ☐ Sill material
- ☐ Insulation
- ☐ Finish materials

1 Frame the Wall. To provide clearance for the beveled windowsill, the wall framing immediately beneath the window must be shorter than it would otherwise be. For a 45-degree bevel, the wall must be shorter by the width of the studs (3½ inches, for example, if you're framing with 2x4 lumber). For a steeper bevel, the wall must be even shorter. Frame out the wall, tip it into place, and fasten the bottom plates to the floor and the top plates to the underside of the joists.

2 Install the Blocking. Cut 1x4 blocking to the width of the window and attach it to the foundation with masonry screws. Don't use nails when you're working this close to the edge of the masonry. Cut a 45-degree bevel on a length of one-by or two-by stock that's the same length as the first piece and nail it to the sill on the framed wall to provide

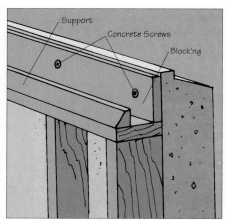

2 Secure blocking to the foundation and nail a beveled support to the wall plate. Bevel the support at the angle you want the sill panel to follow.

support for the sill panel. You can bevel the top edge of the blocking as well, if you like.

3 Install the Sill Panel. The sill panel can be plywood, drywall, or even paneling to match surrounding paneling. In any case, cut a piece to fit beneath the window and tack it temporarily in place. You may have to take out the panel and trim it slightly after the finished wall surfaces have been installed.

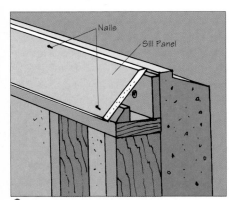

3 Cut the sill panel and tack it into place. Adjust the size later.

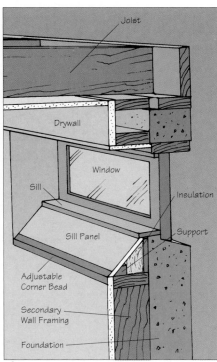

4 Cut the sill panel to finished size, add insulation behind it, and nail the panel into place. Trim the beveled area as needed to cover the edges of the sill panel.

4 Trim Out the Bevel. After installing the finished wall surfaces (usually drywall or plywood paneling), trim the sill panel as necessary for a good fit. Put insulation behind the panel, then nail the panel to the support and the blocking underneath. Add a small horizontal sill to cap the top of the panel. Use corner bead or wood trim to finish off the bottom.

Installing Window Wells

It's possible to fit a small window at the top of a foundation wall and still maintain the mandatory 6 inches above grade (the code minimum). The code is intended to protect wooden building elements from rot by keeping soil away from them. If the windows are too close to the soil, try to lower the nearby grade level. Make sure the ground still slopes away from the foundation. The soil you remove can probably be used elsewhere in the yard. If you can't lower the grade, you'll have to install a window well.

A window well works like a dam to hold soil away from a window that's located partially below grade. Although you can build a well with concrete block, you might find it easier to use a galvanized steel product purchased from a home center. The ribs in a galvanized steel well give it strength, and flanges at each end allow it to be bolted to the foundation walls. Wells come in various sizes. Choose one that's at least 6 inches wider than the window opening and deep enough to extend at least 8 inches below the level of the windowsill.

Difficulty Level:

Tools and Materials

☐ Prefabricated window well
☐ Shovel

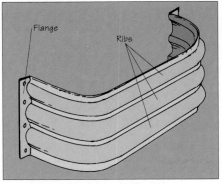

Installing Window Wells. Choose a well that's 6 in. wider than the window opening and deep enough to extend at least 8 in. below the window sill.

☐ Garden hose
☐ Electric hammer drill, masonry bits
☐ Masonry fasteners
☐ Asphaltic mastic
☐ Pea gravel

1 Dig the Hole. Place the well next to the house and use it as a template to mark the perimeter of the hole. Don't try to dig a precise hole. Allow several inches of leeway to maneuver the well into position. Then wash dirt off the exposed portions of the foundation. To allow room for gravel, dig 4 or 5 inches deeper than the depth of the well.

1 Dig a hole that's big enough to contain the window well. Allow several inches of leeway to maneuver the well into position.

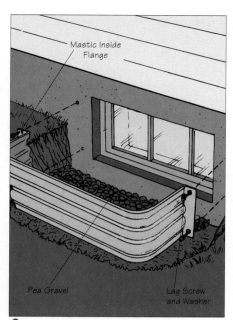

2 *Use the well as a template to mark bolt locations on the foundation, then drill for masonry anchors. After the well is installed, backfill it with pea gravel.*

Remember, the top of the well must be approximately 6 inches above grade.

2 **Attach the Well.** Hold the well against the foundation and mark the position of the mounting holes. Drill the foundation to accept masonry fasteners. Coat the contact areas with asphaltic mastic and install the well. Backfill the outer perimeter with pea gravel, then shovel 4 or 5 inches of pea gravel into the well itself to improve drainage. To keep out accumulations of snow and debris, cover the well with a clear plastic cover.

Skylights and Roof Windows

It's rare for an unfinished attic or garage to provide enough natural light and ventilation for a comfortable, safe living space. Skylights and roof windows are natural partners for windows: Together they can make a dramatic difference in a room, and they're essential for increasing natural light and ventilation. The main difference between skylights and roof windows is that skylights are generally fixed while roof windows are operable for ventilation.

Installing Skylights

The procedure for installing skylights is the same for attics as well as single-story garage conversions. The only differences relate to the depth of the skylight shaft and the placement of the skylights. In a garage conversion, for example, there's no need to consider the placement of a skylight with regard to a horizontal view.

A skylight can be installed in a pitched roof. Some of the work involves cutting through and removing the roof covering. For this reason, the difficulty of the project depends partly on the kind of roof covering on your house. Cutting into a roof covered with slate, clay tiles, concrete tiles, or metal is something most homeowners shouldn't attempt. These materials must be cut with specialized tools, and a skylight installed within such a roof must be waterproofed with special flashing. If your roof is made of anything other than wood shakes or asphaltic shingles, contact a roofing contractor for advice.

Another issue that affects the difficulty of the job is the size of the skylight. Small skylights that fit between existing rafters eliminate the need for special framing. To install larger skylights you have to remove part of at least one rafter, then reinforce roof framing on either side of the roof opening. If you have to remove more than two rafters, consult an engineer for help in determining the proper sizing for headers.

CAUTION: Roof trusses can't be cut, as each part of the truss is an integral part of the entire structure. Exceptions are rare and depend on the style of truss. You must have a truss engineer design a custom alteration before ever cutting any portion of a roof truss. The engineer will determine how the truss must be braced before the job starts and how the truss is to be reinforced around the section that may be cut off.

Choose a Location. The right location for a skylight inevitably is a compromise among aesthetics, function, and ease of installation. Check for nearby tree limbs that may fall or bob in the wind and damage the skylight. Don't forget to consider small trees that may pose a problem in years to come. Prune offending limbs before installing the skylight.

Determine Skylight Height. The higher up on a ceiling you place a ventilating skylight or roof window, the better it ventilates an attic or garage conversion. Keep the unit at least 12 inches from the ridge to provide room for framing and flashing. For attics, placing the unit lower on the ceiling sacrifices some ventilation and privacy but may gain a view.

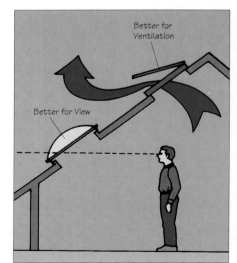

Determine Skylight Height. *The location of a skylight not only affects your ability to reach it but may also determine the window's effectiveness at ventilating.*

Another argument for a somewhat lower placement is to make the unit easy to reach, since you may open and close a ventilating skylight or roof window frequently, depending on the weather. For hard-to-reach units in garages and attic spaces, operating poles are available. Make sure that the unit isn't in the way of cabinetry or furnishings you plan to install in the space.

Determine Horizontal Placement. First you must determine the size of the rough opening required for the particular brand and size skylight you plan to use. The rough opening, sometimes abbreviated RO, is listed in the catalogs you'll use to select your skylight. The rough opening is measured between framing members. Decide approximately where you want the skylight, then adjust the position of the rough opening to the right or left to minimize the need to cut rafters.

Decide on Skylight Wells. As you consider different places for the skylight, determine how you'll trim out the skylight in each instance. Unlike skylights installed elsewhere, such as in garage conversions, attic units don't require much of a light shaft. Instead, the opening beneath the skylight can be boxed off at right angles to the ceiling framing. One variation angles the upper or lower portion of the shaft (or both portions) outward. This reflects more light into the room and isn't difficult to do.

For garage conversions, as well as other household rooms, light shafts will extend from the roof level to the ceiling. All that's needed is a frame made of 2x4s that connects to the rafters at the top and ceiling joists at the bottom. These framing members will simply support the drywall nailed or screwed to them. A bead of white caulking around the frame of the skylight where it meets with drywall will hide any imperfections in the trim.

Types of Skylights

Skylights and roof windows are glazed with either safety glass or plastic. A glass skylight is always flat while plastic skylights (acrylic or polycarbonate) usually are domed. In cold climates both kinds of glazing are best doubled (with an air space in between) to minimize heat loss and reduce condensation problems. Glass is best tempered, laminated with plastic, or wired to improve its strength and minimize danger if it breaks. Color options include tinted glass to reduce glare, reflective glass to limit heat gain in warm climates, and frosted glass for privacy.

Determine Horizontal Placement. *When looking for a place to install the skylight keep in mind that you want to minimize the need to cut rafters.*

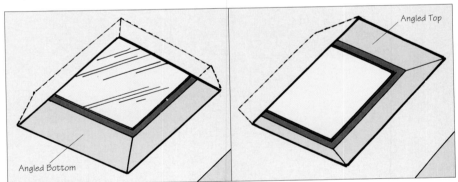

Angled Top

Angled Bottom

Decide on Skylight Wells. *Though the sides of an attic skylight well are flat against the rafters, the bottom (left), the top (right), or both can be angled.*

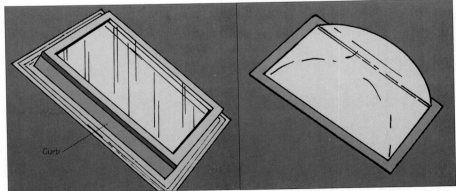

Curb

Types of Skylights. *Flat skylights (left) can be glass or plastic. Domed skylights (right) are always made of plastic.*

To reduce the loss of heat from your house, you can get low-E or argon-filled glazing from some manufacturers. Low-E glass has a coating on one pane that reflects infrared radiation, or radiant heat, back into the room. Argon-filled insulated glazing has the space between the glass panes filled with a dense inert gas like argon instead of air to greatly reduce heat loss through the window due to conduction. If you live in a cold climate, you should consider these options. An inefficient skylight contributes an uncomfortable chill to a room.

Ventilating. There are two basic types of skylights: ventilating and fixed. A ventilating skylight has a hinged flap that can be opened to allow air to flow.

Fixed. Fixed skylights cannot be opened, so they're less complex, less expensive, and somewhat easier to install. If your attic or garage conversion calls for more than one skylight, consider using both types.

Roof Window. A roof window is essentially an operable skylight that you can open fully for ventilation. Roof windows may pivot on the up-slope side or in the middle. They usually have a screen and some sort of mechanism to keep them open.

Roof Window. *This unit opens fully to allow fresh air into the attic.*

Curbed and Curbless. Sometimes skylights are categorized by their method of installation. Older-design skylights rest on a wood frame called a curb that lifts the skylight above the plane of the roof. The curb is usually made of standard lumber and protected from the weather by metal flashing. More-modern self-curbing skylights are attached directly to the roof; no separate curb is necessary. These units are easier to install because they have integral flashing.

Installing a Curbless Skylight

To minimize the chance of damage from inclement weather, plan to complete the basic installation in one day. Once the skylight is in place and tightly sealed, you can complete the rest of the job from indoors at your convenience.

The following steps are for a skylight installation that requires the removal of part of one rafter. If your skylight fits between rafters, the installation is similar but easier. You can use single headers instead of double headers, unless the manufacturer's installation instructions specify otherwise.

Difficulty Level: 🔨🔨🔨

Tools and Materials

- ☐ Goggles
- ☐ Gloves
- ☐ Basic carpentry tools
- ☐ Framing lumber
- ☐ Spade
- ☐ Chalkline
- ☐ Circular saw with carbide blade
- ☐ Galvanized common nails, 12d, 16d
- ☐ Roofing nails, 1½-inch
- ☐ Caulking gun and silicone caulk
- ☐ Roof shingles to match existing shingles

1 Locate the Roof Opening. Lay out the rough opening on the underside of the roof sheathing.

Drive a nail through the sheathing (and the roofing) at each corner of the layout. The nails help locate the layout when you cut the sheathing from outside while on the roof. To strengthen the roof around a large skylight, leave room for an extra rafter, called a trimmer, on each side.

2 Remove the Shingles and Cut the Roof Opening. Use an ordinary garden spade to pop up as many courses of shingles as necessary to clear the general area in which the skylight will be located. Remove all nails and use a utility knife to trim away the shingles and the underlying roofing felt from what will be the opening. The amount

1 *Nail through the underside of the sheathing to mark the cutout corners.*

2 *Be sure you have a firm footing before cutting the sheathing.*

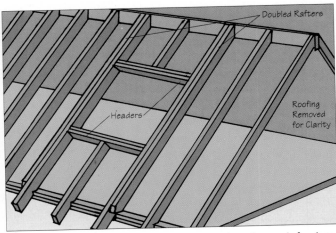

3 *Use a combination square to mark the rafter cut lines. Use 2x4s or 2x6s as temporary supports for rafters to be cut.*

4 *Use 16d nails to install the headers and to sister reinforcing rafters to each side of the existing rafters.*

trimmed depends on the type of flashing used. Some curb-type sky-lights call for multiple courses of flashing, called step flashing, along the sides of the curb, while others use a piece of flashing called a "collar" to keep out the water. When using step flashing, trim the roofing about 2 inches away from the opening. If you're using one-piece flashing, trim enough roofing away so that the flashing rests directly on the sheathing. In either case, the goal is to remove as little roofing as possible.

Snap chalk lines between the nails driven through the roof in Step 1 to mark the portion of the sheathing to be removed. Use a circular saw set to a depth of approximately ¾ inch. Don't cut through the rafters. If the roof is pitched steeply, it may be bet-ter to use a reciprocating saw from inside the attic. After the cut is com-plete, remove sawdust and debris from the roof to keep from slipping on it.

3 **Cut the Roof Framing.** Use a combination square to mark cut lines along the sides of the rafter to be cut. The top and bottom cuts will be 3 inches outside the rough open-ing to provide room for doubled top and bottom headers. To pick up the roof load of the severed rafter, nail

5 *Prepare the rough opening and set the skylight unit in place. Secure the skylight with the hardware provided.*

6 *Replace shingles around the skylight to complete the installation. Don't nail through the bottom skylight flange.*

braces across several adjacent raf-ters and remove them only after the headers are in place. Use a cir-cular saw to cut the rafter partway through. Then use a handsaw to finish the cut.

4 **Install the New Framing.** Slip the headers into place and use 16d nails for each connection through the rafters. Install support rafters on both sides of the rough opening and nail them to the existing rafters. In order to be effective, the bottom of the support rafter must

rest on the wall plate and its top must fit against the ridge.

5 **Place the Skylight.** Place the skylight over the opening and secure only the top and side flanges. You may have to caulk the perime-ter of the opening before setting the skylight in place. Use 1½-inch roof-ing nails.

6 **Complete the Installation.** Extend the roof shingles over the top and side flanges of the sky-light and under the bottom flange.

Wiring and Plumbing

8

All attic, basement, and garage conversions have to be wired for outlets and lighting. If you're adding a home office, you may need to accommodate other electrical appliances like computers. And if you're building a master bedroom suite or including a new bathroom in the conversion, you'll also face the challenges of adding plumbing.

Wiring

Whatever the scope of your project, it has to be wired and lighted. In some cases, such as building a home office, the electrical requirements may be considerable. Once you understand some of the basic concepts, wiring isn't that difficult. It does require, however, that you pay stringent attention to safety and electrical codes. In some parts of the country only licensed electricians are allowed to work on household wiring, while in other locales a homeowner may do all the work on his or her own home as long as the finished project is reviewed and approved by an electrical or fire inspector. Be sure to check local codes before beginning work.

Additional Circuits. Although it's possible to extend a circuit to supply electricity to an attic, basement, or garage conversion that will have modest needs, doing so may overload the circuit. Some codes, in fact, require that the renovated space be equipped with new circuits. Not only are new circuits safer, they also make the conversion far more convenient to use.

Service Entrance. Electricity enters the house through a meter that measures the amount of electricity used. It then enters the service entrance panel. The panel is essentially a distribution center that sends incoming electricity to various portions of the house. Each circuit is protected by a fuse or a circuit breaker that cuts power to a circuit in the event of an overload or circuit fault. Each circuit is independent of the others, so when power is cut to one, the others remain unaffected and continue to do their jobs.

To add one or more circuits you must route wire to the new living space, connect all the outlets and switches

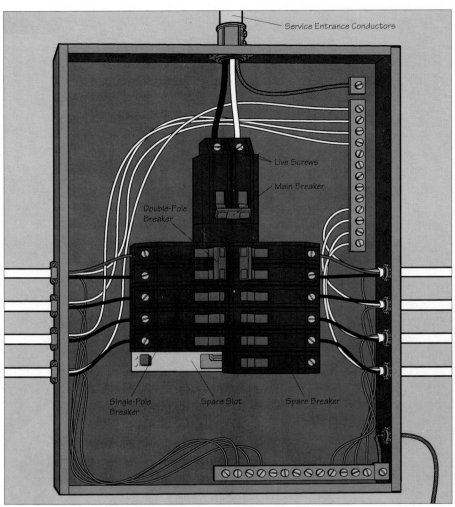

Service Entrance. *Each incoming cable is connected to a circuit breaker. This panel has room for two additional breakers, so it can serve up to two more circuits.*

to the new circuit, cut power to the service panel, and add the circuit breakers. If you're not familiar with this work, turn it over to a licensed electrician. To save money, however, you can do the time-consuming task of routing the wiring. Just leave the connections to the electrician.

Tools and Materials. Virtually all wiring jobs can be accomplished using the small assortment of tools mentioned at the beginning of the book, on page 7. For running wire up to the attic, down to the basement, or out to the garage through finished walls, a fish tape is indispensable. Other essential tools are shown in "Tools and Materials" on the facing page.

Electrical Safety

■ Always turn off the power at the main electrical service panel before beginning work.

■ Always use tools that have insulated handles. Don't use screwdrivers that have metal shanks that extend completely through the handle. Even though the handle is insulated, the exposed shank can transmit an electrical shock to your hand.

■ Never use a metal ladder when working with electricity. Use a wood or fiberglass ladder instead.

■ Always use a voltage tester to test a wire for the presence of electricity before you work on it (even if you switched off the circuit).

Running Cables

House circuits are generally wired with nonmetallic sheathed cable, often referred to by the trade name Romex. The cable, which is flexible and easy to work with, is made of two or more copper wires wrapped in a protective plastic sheathing and sold in rolls of 25, 50, 100, or 250 feet. Aluminum wire was used widely from World War II through the mid-1970s but is no longer considered suitable for household wiring. Consult an electrician before modifying an aluminum wire system in any way.

As it loops through the house, wiring is supported by heavy-duty cable staples that are driven into framing lumber with a hammer. Make sure the staples are sized to the specific gauge of wire used.

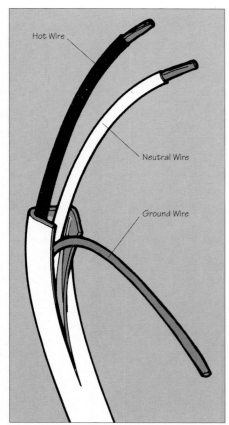

Running Cables. *Nonmetallic sheathed cable is the standard for residential wiring. The cable has a sheathing of flexible plastic that protects several wires.*

Hot Wire

Neutral Wire

Ground Wire

Joining Wires

At one time, all wires in a household system were spliced together with solder and electrical tape. Now, splices are made by joining wires with plastic caps called wire connectors. The inner portion of each cap is threaded. To join two or more wires, simply strip insulation from each one, hold them together, and twist the wire connector in a clockwise direction.

Tools and Materials

■ **Screwdrivers.** You need a flat-bladed screwdriver and a Phillips screwdriver, each with a nonconducting handle.

■ **Needle-Nose Pliers.** This is the perfect tool for snipping a wire to length and bending the end into a tight loop to go around terminal screws.

■ **Wire Stripper.** This tool has cutter holes with diameters to match various wire gauges. It easily strips off insulation without nicking the wire itself.

■ **Fish Tape.** Essentially a roll of stiff wire, the fish tape has a hook on one end to which wire cable is attached. The wire is then pulled through the wall.

■ **Voltage Tester.** With the probes inserted as shown, this inexpensive tool is used to test for the presence of electricity.

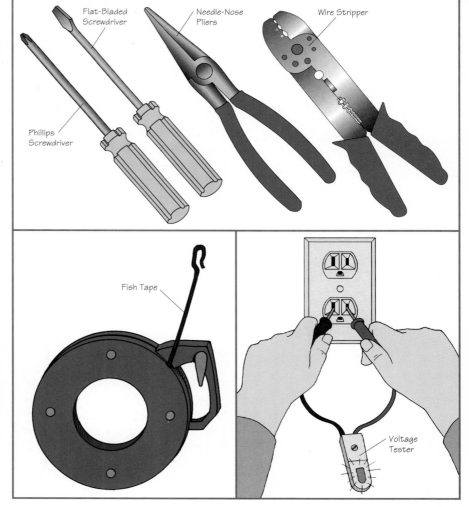

Flat-Bladed Screwdriver

Needle-Nose Pliers

Wire Stripper

Phillips Screwdriver

Fish Tape

Voltage Tester

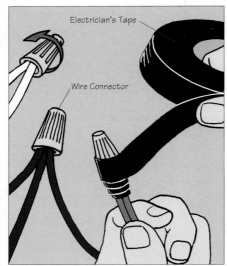

Joining Wires. Wire connectors are small plastic caps that join the ends of wires. Even though the caps should stay in place on their own, it's a good idea to wrap the base with electrician's tape.

Connectors come in many sizes; choosing the right one depends on the number of wires to be joined and the gauge of the wires. Most often you'll join two or three No. 14 or No. 12 wires. It's cost-effective to buy a box of wire connectors in the size most often needed rather than a few at a time. Have some plastic electrician's tape on hand. Electricians often wrap a turn or two of tape around the base of a wire connector to ensure its staying power. Tape is useful for other purposes as well.

Choosing the Right Cable

The individual wires (called conductors) in a cable are available in a range of diameters. These diameters are expressed in gauge numbers; the higher the gauge number the smaller the wire diameter. The more amperes a circuit is designed to carry, the larger the wire diameter requirement. Amperage is a measure of current flow. Circuits serving lighting and standard receptacles typically are 20 or 15 amps. Use 12-gauge wire for 20-amp circuits and 14-gauge

wire for 15-amp circuits. Markings found on the plastic sheathing of cable explain what's inside and identify the kind of insulation covering.

Consider the following designation, for example: 14/2 WITH GROUND, TYPE NM, 600V (UL). The first number tells the size of the wire inside the cable (14 gauge). The second number tells you that there are two conductors in the cable. There's also an equipment grounding wire, as indicated. Each wire is wrapped in its own plastic insulating sheath, though the ground wire is most likely bare. In this case, the type designation indicates a cable that's for use only in dry locations, in other words, indoors. Following the type is a number that indicates the maximum voltage allowed through the cable. Finally, the UL (Underwriters Laboratories) notation assures you that the cable has been certified as safe for the uses for which it was designated. For safety reasons, never use wiring or other electrical supplies that don't bear the UL listing.

Estimating Wire Needs

New wiring that leads from the service panel to the first switch or outlet in the attic, basement, or garage conversion must be a continuous and unbroken length. The code allows certain exceptions to this rule, but a single length is ideal and almost always possible. You may have to snake wiring over, around, or through a number of obstructions en route to the attic, basement, or garage conversion, particularly if the service panel is a long way from those areas. Normally, service panels are relatively close to, or in, a basement or garage. Rather than try to calculate the length of this path, begin with a 50-foot roll of wiring, which in most cases is more than enough to reach from a panel to an attic or other area. The

excess, if any, can be used for general needs once the attic, basement, or garage conversion is reached.

When running wire through a structure in which all the walls are exposed, such as an unfinished attic, it's fairly easy to figure out how much you need. Measure the distance between each connection you have to make and add a foot for every connection. Total this amount, then add 20 percent to provide for a margin of error.

Wiring the Attic

Getting wire from the service panel to the attic calls for considerable ingenuity. Every house is different, but the following guide provides some techniques for solving typical problems. Don't connect wires to a power source until they've been safely connected to outlets or fixtures.

Holes in Framing. Drill a ⅝- to ¾-inch diameter hole in places where cable must pass through the framing.

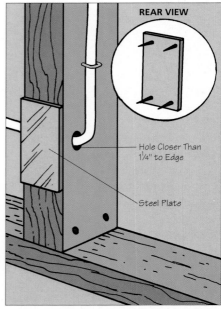

Holes in Framing. A nailing plate protects cable that passes close to the edge of a joist or stud. You pound in the barbs on the plate with a hammer.

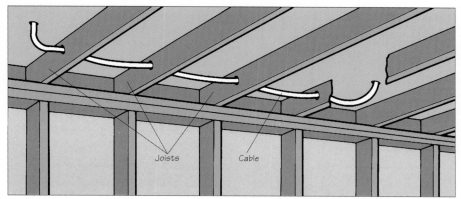

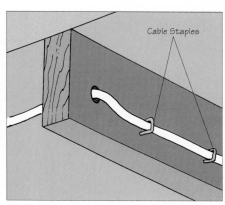

Holes in Joists. *Use an auger bit and an electric drill to bore through joists. Locate the holes to minimize joist damage and keep the cable out of harm's way.*

Cable Support. *Use heavy metal staples to hold cable in place without pinching it.*

Though you can use a spade bit to do this, an auger bit (available at many hardware and home supply stores) is easier and safer to use. The hole must be located at least 1¼ inches from the edge of the stud to minimize the risk that a paneling or drywall nail or screw will contact the wire at some point. If for some reason the hole must be closer, the National Electrical Code requires that the wiring be protected with a steel plate.

Holes in Joists. When drilling through a joist, always put the hole in the center one-third of the board. The joist will be weakened if you drill through the bottom third, and you may run into nails if you drill through the upper third. Avoid notching the bottom of a joist to create a path for wiring. If direct access to the attic doesn't exist—and in most cases it won't—try to route wiring along forced-air ducts, through pipe chases, or even through voids between the framing and a masonry chimney. Plumbing waste vents always lead to the roof, so look for a path along one of them that might lead to the attic. In some cases you may have to run wiring into the garage, up the inside of a garage wall, then into the attic. Often, exposed wiring must be housed in rigid electrical conduit that's properly anchored to the wall. Check your local codes.

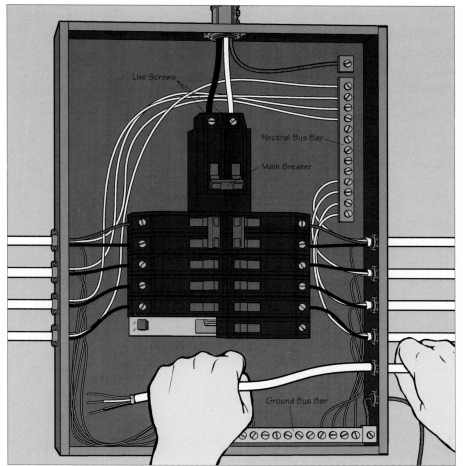

Panel Wiring. *When stringing cable to new circuits, leave extra length in or near the box to connect the cable to the proper circuit breaker and bus bars.*

Cable Support. According to code, you must use cable staples to support the cable (generally at least every 54 inches on a run and within 8 inches of boxes). Be careful not to damage the outer casing of the wire as you drive the staples home.

Panel Wiring. Make sure you leave at least a 24-inch tail of cable at or in the service panel. There must be enough cable for the electrician to connect wire to the circuit breaker and bus bars.

1 *From the basement, search for the wall through which you wish to run wiring. Look for clues, particularly other cables running through the floor.*

Running Cable through Finished Walls

If you can't find a clear path to the attic, you may have to run wire through existing interior walls. Use two fish tapes and a helper to guide the wire through wall cavities.

Difficulty Level:

Tools and Materials

- ☐ Goggles
- ☐ Electric drill with sundry bits
- ☐ Utility knife
- ☐ Keyhole saw
- ☐ Fish tape
- ☐ Electrical tape
- ☐ Electrical cable
- ☐ Metal plates for studs

1 **Locate the Bottom Plate.** From the basement, locate the underside of a wall through which you're going to run wire. Look for nails that penetrate the subfloor as a clue to the location of the wall. Wiring that disappears into a hole in the subfloor is another clue. You can also measure from an exterior wall, or from a landmark that's visible above and below the floor (a stairway or plumbing pipe, for example). Identify all doorways in the area. If you're running wire to the attic of a two-story house, make sure there's a second-story wall directly over the first-story wall you've chosen.

2 **Drill an Access Hole.** Once you've located the wall, drill a ¼-inch pilot hole through the floor directly below the wall. Stick a length of coat hanger through the hole, securing it temporarily, then go upstairs. If you can't see the coat hanger, you've drilled successfully into the interior of a wall. Go back and enlarge the hole using a ⅝- to ¾-inch bit.

3 **Locate the Top Plate.** From the attic of a one-story house, locate the top plate of the same wall and drill a ⅝-inch hole through it. The hole must be above the first hole.

4 **Make the Ceiling Notch.** If you're working in a two-story house, you'll need to pull the wire through in two stages. The first

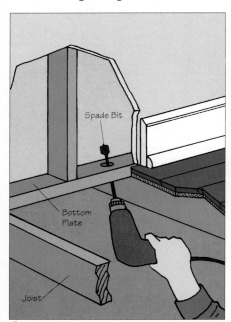

2 *Drill a small test hole in case you misjudged the wall location. Enlarge the hole when you're sure it's right.*

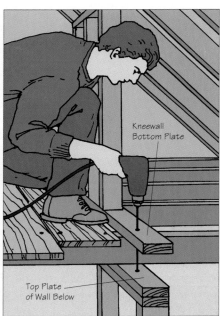

3 *From the attic, drill through the top plate in a spot directly above the hole in the bottom plate of the same wall.*

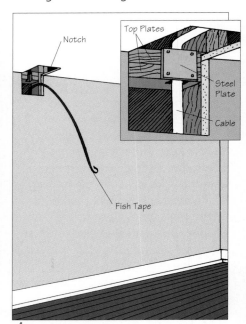

4 *For a two-story house, make a notch in the first-floor ceiling so you can pull cable in two stages.*

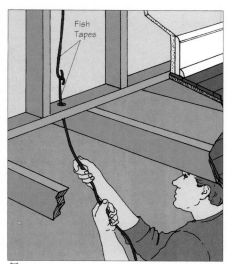

5 *Slip fish tape into the wall from both directions and hook them. If you encounter blockage in part of the wall, drill another set of holes elsewhere.*

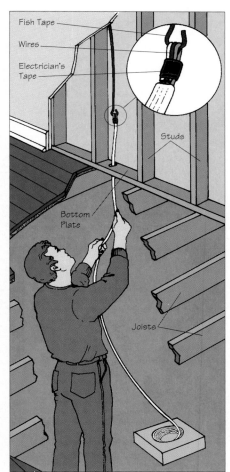

6 *Pull the upper tape into the basement and secure cable to it with electrician's tape. Make sure that nothing snags the wires as a helper pulls the cable through the wall.*

stage is getting the wire from the basement to the first-floor ceiling. To run the wire, make a small cutout at the ceiling-and-wall juncture to expose the first-floor top plate. Notch the plate so you can get the cable into the second-story wall.

5 **Fish for the Wire.** For a one-story house, slide fish tape into the wall through the hole in the basement and the hole in the attic and try to hook them together. This isn't easy, particularly if there's wiring or pipes in the wall already. For a two-story house, hook tapes from the basement and the ceiling notch.

6 **Attach the Cable.** Pull the snagged upper tape into the basement, attach the cable to it, and tape them together. For a one-story house, hoist the upper tape through both holes and into the attic as a helper feeds wire into the wall. For a two-story house, hoist the cable through the first-floor ceiling notch. Disconnect the cable, hook fish tapes from the notch and the attic hole, and repeat the fishing process to bring the cable into the attic.

Installing Receptacles

New receptacles accepted by the National Electrical Code contain three slots: two vertical slots of slightly different length for the hot and neutral wires and a U-shaped slot for the ground wire. You'll need special ground-fault circuit-interrupter (GFCI) receptacles for basement areas, bathrooms, and other areas near water or below ground. Like a supersensitive circuit breaker, a GFCI cuts current in a fraction of a second if a short occurs. Check local codes for specifics.

You can wire a regular outlet using the screws on the sides and wrapping the receptacle with electrician's tape or, more simply, by inserting the stripped end of a wire into the proper hole in the back. Some electricians don't consider the holes to be as reliable as screw attachments. GFCI outlets have pigtails that you attach to the circuit cable.

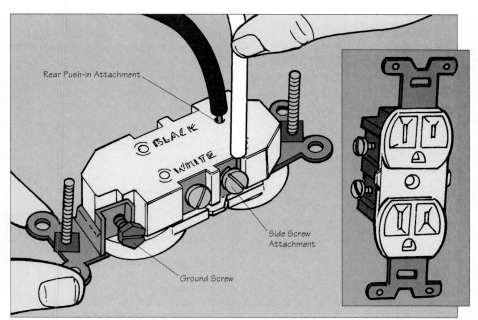

Installing Receptacles. *The currently accepted receptacle design for 120-volt circuits has two vertical slots of slightly different length for the hot and neutral wires and a U-shaped slot for the ground wire. Wire may be attached at the sides or rear.*

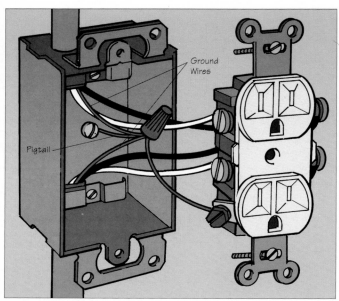

Wiring Middle-of-the-Run Receptacles. *Connect the two black wires to the two brass-colored screws and the two white wires to the two silver-colored terminals.*

Wiring End-of-the-Run Receptacles. *Bring the incoming cable into the box. Connect the black wire to a brass-colored screw and the white wire to a silver-colored screw.*

Wiring Middle-of-the-Run Receptacles

Bring the incoming and outgoing cables into the box through the top and/or bottom holes. Connect the two hot (black) wires to the two brass-colored screws (or into the holes marked "black" on the backside). Attach the two neutral (white) wires to the silver-colored screws (or into the holes marked "white"). Connect the ground wires together and to the grounding screw in the box (if metal).

Wiring End-of-the-Run Receptacles

Bring the incoming cable into the box through the top or bottom hole. Connect the hot (black) wire to a brass-colored screw (or into the hole marked "black" on the backside). Attach the neutral (white) wire to the silver-colored terminal (or into the hole marked "white"). Connect the ground wire to the grounding screw in the box (if metal).

Wiring Fixtures

Wiring lighting fixtures or any electrical appliance in a permanent location requires bringing a power supply cable to the fixture, wiring in a switch, and attaching the fixture to the wall or ceiling. You can wire a remote switch by running the power cable through the switch and into the fixture (in-line wiring) or by running the power cable to the fixture first, then taking a "leg" off the hot wire to the switch. Here's how to do both for a wall or ceiling lamp. Use the same steps for any fixture, but be sure to read the manufacturer's instructions. Remember always to shut off power to the circuit at the main breaker before beginning any wiring job.

Difficulty Level:

Tools and Materials

☐ Basic electrical tools
☐ Cable, 14-gauge
☐ Plastic wire connectors
☐ Lighting fixture
☐ Switch

Wiring a Fixture in Line

1 **Bring a Cable to the Switch.** Bring a power cable from a junction box or receptacle into the switch box. Run an outgoing cable from the switch box into the fixture box.

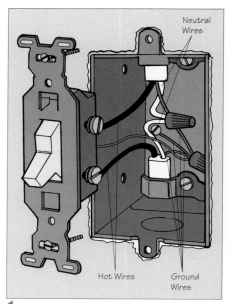

1 *Bring a power cable from the power source (junction box or receptacle) into the switch box. Then run an outgoing cable from the switch box into the fixture box.*

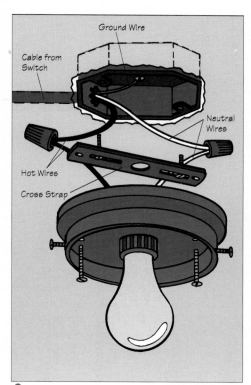

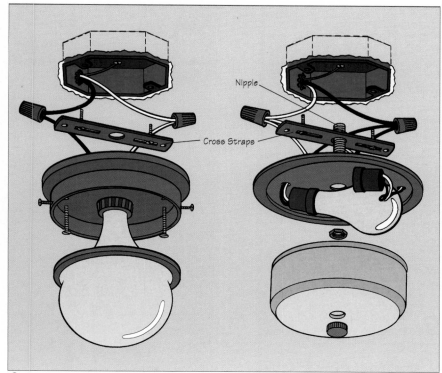

2 *Bring the cable from the switch into the fixture box and connect the wires.*

3 *Attach incandescent ceiling and wall lamps to the cross strap with screws (left) or to a nipple screwed into the strap (right).*

Connect the hot (black) wire from the power cable to the side of the switch marked "hot" or "black" (use the screw or push-in hole). Connect the black wire of the outgoing cable to the other hot terminal. Splice the two neutral (white) wires with a plastic wire connector. Splice the grounding (bare) wires with a plastic wire connector and attach a short length to the green grounding screw inside the box (if metal). Insert the connected switch into the box and attach the screws and cover plate.

2 **Connect the Cable to the Fixture.** Bring the cable from the switch into the fixture box and attach the black and white cable wires to fixture wires of the same color with plastic wire connectors. Connect the ground wire to the box (if metal). In plastic boxes, connect the ground wire from the fixture to the ground wire of the cable, if the fixture has a ground wire. Otherwise, do nothing with the bare wire.

3 **Install the Fixture.** Fixtures attach to electrical boxes in various ways. Two common methods for attaching incandescent ceiling and wall lamps are shown above. Begin by screwing a cross strap across the box. If the fixture base has screws at the sides, position the base so that these screws align with the two holes in the cross strap. Insert the screws and install the lamp and lens. If the lamp has a nipple in the center (instead of side screws), screw the nipple into the center hole in the cross strap of the box. Then place the base and lens over the nipple and install the nut to hold the parts together.

Wiring a Fixture Switched from a Loop

Sometimes it's not feasible or convenient to wire the fixture in line. In that case, you can run the power cable directly to the fixture and run a loop cable from the fixture to the switch.

1 **Run the Cables.** Bring a power cable from a source, such as a receptacle or junction box, into the fixture box and strip the wire ends. Run a separate loop cable from the fixture box into the switch box and strip the wire ends.

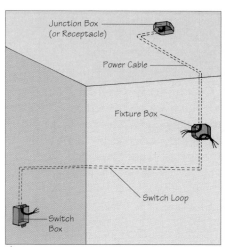

1 *Bring a power cable from a source, such as a receptacle or junction box, into the fixture box, then run a switch-loop cable from the fixture box into the switch box.*

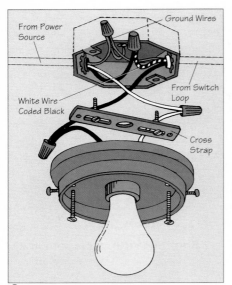

2 To wire a lamp controlled by a switch loop, connect the white power wire to the white fixture wire and the black power wire to the marked white loop wire.

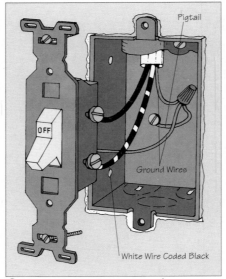

3 Connect the wires from the switch-loop cable to the screws in the side of the switch. Insert the switch into the box, and tighten the screws.

2 Connect the Power Cable.

Connect the white wire from the power cable to the white lead of the fixture. Connect the black fixture wire to the black wire of the switch-loop cable. Then connect the white wire from the switch loop to the black power lead. Mark the end of this wire as "black" by wrapping it with electrical tape or coloring it with a marker (to prevent it from being confused for a neutral wire). Finally, connect the ground wire from the power cable to the ground wire of the switch loop and to the grounding screw on the box (if metal).

3 Connect the Switch.

Connect the black and white wires from the switch-loop cable to the screws in the side (or holes in the back) of the switch. Code the white wire as "black" by wrapping the end with electrical tape or coloring it with a marker. Connect the ground wire to the grounding screw in the box (if metal). Insert the switch into the box and tighten the screws.

Install the fixture base and lens on the box as described for "Wiring a Fixture in Line," page 146.

Wiring a Basement

Building codes require that basements be supplied with a minimum of one circuit, though at least one additional circuit will make the basement far more convenient to use. A home office has considerable electrical requirements, so plan at least one circuit for this room alone.

Wiring Surface-Mount Systems

It's difficult to run wiring and install boxes on masonry walls. It may be easier to build secondary walls against the foundation and run wire through them instead. If wiring directly onto the masonry walls makes sense, however, be sure to check local codes, particularly when it comes to grounding the metal parts of the system.

You can route wiring along the surface of a solid concrete or concrete block wall as long as it's contained in a system that protects the wires

from mechanical damage. One system calls for the use of metal boxes and thin-wall metal conduit called EMT, for Electrical Metal Tubing. Special connectors form a tight seal between the box and conduit. Conduit is secured with metal straps that you attach to the masonry with screws. Electrical boxes are also screwed in place. Installing the system calls for special tools, and because sheathed cable is too bulky, you'll have to pull individual wires through the conduit.

A similar surface-mount system is easier for do-it-yourselfers to install. Instead of metal conduit, it uses a plastic track with a snap-on cover to contain the wire, along with a series of fittings for changing direction and splicing lengths of track together. The only special tool you'll need is a hacksaw. Check local codes before installing this kind of system.

The system can be purchased at home center stores. Make a sketch of the basement that shows approximately where switches, outlets, and fixture boxes will be installed, and

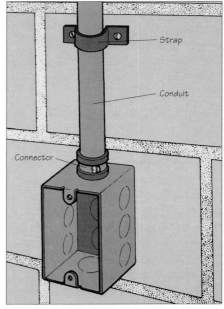

Wiring Surface-Mount Systems. *Route individual conductors through lightweight metal tubing that you clamp to each electrical box.*

take it shopping with you. If the supplier allows unused materials to be returned, purchase more supplies than you think are necessary, so you don't get caught short.

Difficulty Level:

Tools and Materials

- ☐ Basic electrical tools
- ☐ Electric drill and assorted masonry bits
- ☐ Masonry fasteners
- ☐ Surface-mount track, fittings, and boxes
- ☐ Electrical cable (type THHN)

1 Install the Power Feed. Surface-mount systems use single conductors rather than sheathed cable. You can use sheathed cable to connect the surface-mount system to the ser-vice panel, though. A special adapter plate that fits over a standard electrical box makes the transition from sheathed cable to individual conductors. The surface-mount box contains cutouts in all four sides of the box to accommodate track that goes up, down, or to the sides.

2 Install the Channels and Elbows. Begin at the starter box and install sections of base track, available in 60-inch lengths. Drill a hole through the base track every 18 inches and ½ inch from each end, then use it as a template for marking holes on the wall. Drill holes in the wall for masonry screws or plastic shields, then screw the base track to the wall. Be careful not to tighten screws so much that they damage the track. Route the base track to the area of each switch and outlet box, and secure the box with masonry screws or plastic shields according to the manufacturer's instructions.

3 Install the Intersections. In the places where lengths of base track intersect mid-run, cut away the lip of the track to make room for wires. Tracks that intersect at an inside or outside corner are butted together. Turns on the same wall are mitered. Use a hacksaw and a miter box to make the cuts.

4 Run the Wires. Surface-mount systems use type THHN conductors instead of sheathed cable. As many as ten 14-gauge conductors or up to seven 12-gauge conductors can fit in a channel. Use plastic clips to hold wires that stretch from switches to boxes.

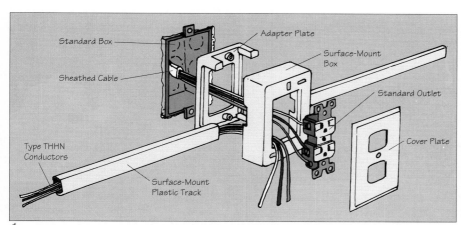

1 Attach an adapter plate to the existing box. Then install a surface-mount box and connect the surface-mount track to it.

2 Drill holes in the base track no greater than 18 in. apart.

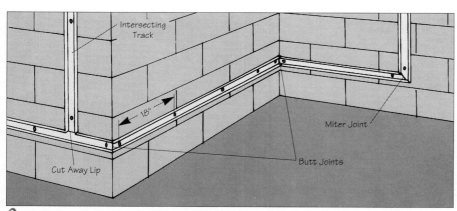

3 For mid-run intersections, cut away the lip as shown. Use butt joints for inside and outside corners, miters for turns on the same wall.

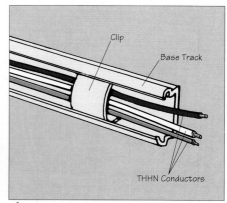

4 Route individual conductors to each switch, outlet, and fixture.

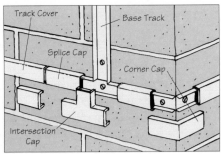

Track Cover · *Base Track* · *Splice Cap* · *Corner Cap* · *Intersection Cap*

5 *Covers snap over the base track while joint caps snap over intersections and seams. Conductors are not visible when the installation is complete.*

5 **Cap the Base Track.** Cut lengths of base-track cover to fit over the base track and snap them in place. Cut covers 1⅜ inches short of each intersection to accommodate the various joint caps.

Relocating Existing Wiring

In most cases, a great deal of wiring already exists in the basement. These wires feed circuits elsewhere in the house and may have to be moved depending on where they are and what's planned for the basement ceiling.

If wires run through holes in the joists, there's no need to worry about them unless they come closer than 1¼ inches to the edge of a joist. If so, nail a protective metal plate to the joist to prevent the wires from being punctured by nails. If a suspended ceiling is to be installed, the wires don't have to be relocated. However, if a drywall ceiling will be installed and wires run along the underside of the joists, the wires must be relocated.

Difficulty Level:

Tools and Materials

☐ Basic electrical tools
☐ Combination square
☐ Saber saw (or handsaw)
☐ Metal plates for joists

1 **Loosen the Wires.** Use nippers to grasp the edge of each cable staple, then lever out the staple. As you pull the staples, don't crush the cable beneath the nippers and don't nick the outer casing of the cable. Move the wires aside temporarily and dispose of the staples.

2 **Cut a Notch.** According to building codes, notches in the bottom of a joist must be no more than one-sixth the depth of the joist, and they must not be located in the middle third of a joist's length. If these criteria can't be met, disconnect the circuit and run the wires through holes in the joists. To lay out the notch, set a combination square to the depth of the notch (just deep enough to contain the wires) and mark cut lines on the edge of each joist. Use a saber saw or handsaw to cut both sides of the notch.

3 **Complete the Notch.** The bottom of the notch will be parallel to the grain of the joist, so you can knock out the waste wood easily by striking it with a hammer. If necessary, use a chisel to clean up the bottom of the notch.

4 **Relocate the Wires.** Move the wires into the notch. If necessary, use a cable staple to hold them. According to electrical codes, the

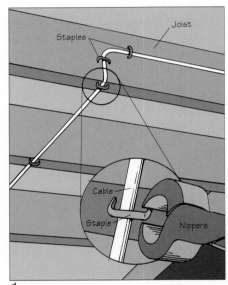

Joist · Staples · Cable · Staple · Nippers

1 *To remove a cable staple, grasp one side with nippers and pull out the staple. Don't damage the cable itself.*

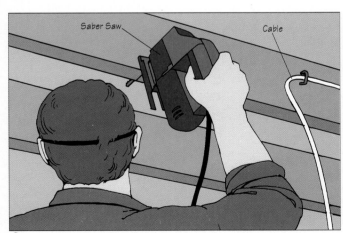

Saber Saw · Cable

2 *Never notch a joist in the middle third of its length. Use a saber saw or handsaw to make the shallow cuts.*

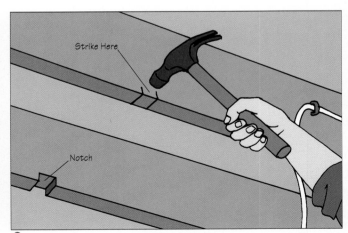

Strike Here · Notch

3 *Strike the notched area with a hammer to knock that chunk of wood out. Clean up the notches with a chisel.*

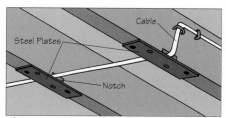

4 *Move wires into the notches, and nail protective metal plates over the notches and into the edges of the joists.*

wires must be protected by a steel plate that's at least $\frac{1}{16}$ inch thick.

Wiring a Garage

Perhaps the easiest room conversion to wire is a garage. This is because in many cases electrical service panels are located on a garage wall. The reason is two-fold: First, home garages are generally located at the front of a house, which makes it easy for electric company personnel to find and read the meter. Second, service panels are not aesthetically pleasing to the eye; they're more or less "hidden" in a garage rather than

being placed somewhere more visible in the house. Also, service panels are usually placed just on the other side of the wall from meters.

Garages usually have open attic spaces, and their walls may not be covered with drywall. The one exception is walls that separate living space from the garage. According to code, those walls must be covered with $\frac{5}{8}$-inch drywall as a means of added fire protection for the living space. With a service panel close by, an open attic space, and open wall framing, you should be able to run wire unobstructed. Keep in mind that partition walls will also need outlets and switches. So, do all of the framing work before running wire.

Plumbing

As with wiring, plumbing pipes must be installed before walls, ceilings, or floors are covered. A household plumbing system consists of two parts: A fresh water supply (hot and cold water) furnished through copper tubing by way of a well or municipal water system, and a drainage network designed to carry waste away from the house to a municipal sewer or private septic system. Although related, each plumbing system is independent. Pumps force fresh water through tubing under controlled pressure, so the tubing can be installed at any angle and in

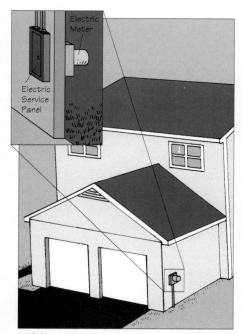

Wiring a Garage. *It's common to find an electric meter at the front of a garage so it's easy for meter-readers to see and read.*

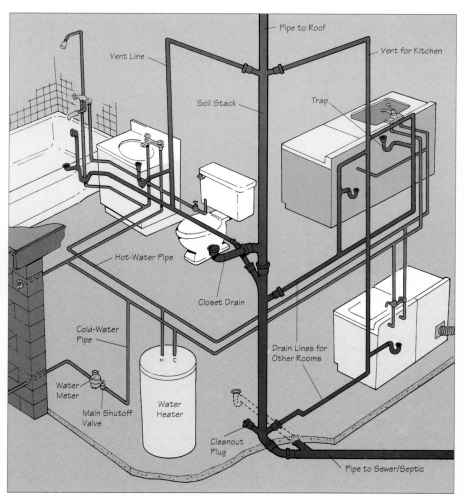

Plumbing. *Every plumbing system should have: (A) pipes that deliver water, (B) pipes that transport soil/waste to a main drainpipe, which empties into a sewer or septic system, (C) water-filled traps that keep gases out, (D) vent pipes, and (E) cleanout plugs.*

any direction. Waste is carried to septic systems or sewers through gravity flow, so waste pipes must run downhill from fixtures to main sewer or septic lines at a slope of at least ¼ inch per foot.

In addition, waste pipes (commonly referred to as the drain-waste-vent, or DWV, system) must be vented to the atmosphere so air pressure within the network of DWV pipes is equalized. Vents prevent airlocks so water can drain freely and prevent water in plumbing traps from being siphoned out. Every plumbing fixture and plumbing appliance in a house has a trap that should stay filled with water to block the passage of sewer or septic-tank gases into the house.

Bathroom installation in any attic, basement, or garage conversion demands thorough planning. Walls must be opened up to gain access to existing plumbing pipes and new routes of travel must be determined for the new bathroom's piping network. Ideally, a new bathroom should be constructed directly above, just below, or adjacent to an existing bathroom. This way, existing plumbing pipes are close by and relatively easy to access, especially the DWV pipes that have to be vented through the roof: New DWV pipes can be tied into nearby existing vent stacks. Plumbing for a new bathroom must be done under permit, and a plumbing inspector will have to test the new pipes to ensure they don't leak. If your plumbing experience is limited, you'll be best off hiring a professional plumbing contractor to complete the work.

Attic Bathrooms. Installing a new bathroom in an attic conversion should be relatively easy, especially if the floor joists are 2x6s or larger and an existing bathroom is located just below. The space between the floor joists offers an ideal location for placing drainpipes. Plan to construct one of the bathroom partition

Attic Bathrooms. *If possible, plan the bathroom so you can connect the toilet to the main soil stack and run drain lines between joists.*

walls directly above and in line with the existing bathroom's soil stack, the pipe that probably runs up through the attic and then through the roof. The wall containing the soil stack should be framed with 2x6s so there's enough room inside the wall cavity to conceal the 3-inch main stack pipe.

Basement Bathrooms. For those experienced in cutting and soldering pipe, the job of adding tubing to supply water to a basement lavatory or toilet is straightforward. Providing a drain and vent for the waste water, however, remains a difficult job, especially if the basement is below the level of the existing DWV system, as it often

is. Getting waste to a main sewer or septic line above the basement floor may require the installation of a pump system. Contact a plumber before finalizing a bathroom arrangement in the basement.

Garage Bathrooms. As with an attic or basement, running fresh water to a new bathroom in a garage conversion is easy; the tough part focuses around the drainpipe. Ideally, a DWV stack will be located in a garage wall to serve a bathroom on the other side. Plan to build your new bathroom in that area to take advantage of the proximity of the drain and vent lines. Remember, however, that a toilet, shower, or bathtub drain will have to be positioned under the

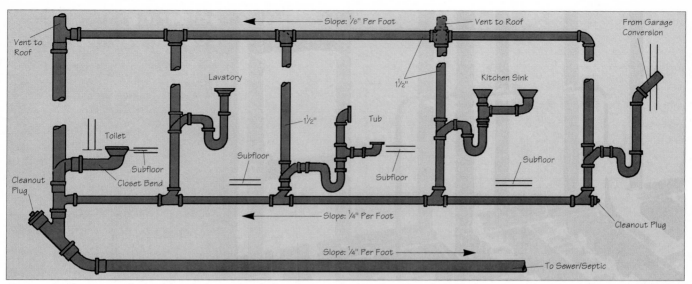

Garage Bathrooms. *All plumbing drain systems in houses must be big enough to handle the expected volume and must vent to the atmosphere to keep the system equally pressure balanced.*

floor. You'll either have to cut a section of concrete out for the installation of drainpipes or install a platform floor for the bathroom.

Getting Started

During the planning stage for your new bathroom, draw up a detailed plan to scale. Once the plan has been prepared, along with a list of parts and materials, have someone with experience in doing a similar project check to see whether any-

thing has been overlooked. The plumbing supply store with which you do business may have trained consultants on staff who can provide such a service.

1 Make a Plan. After deciding where new fixtures and appliances are to go, make plan-view and elevation drawings. In planning the layout, leave ample clearance between fixtures. Get a copy of your local plumbing code from the building department. You'll find informa-

tion regarding sizing and slope of pipes, venting methods, cleanout plug placement, and the like.

Make notes on your plan of the type of water-delivery and DWV pipes already in your home. Consult your local code to determine whether new pipes and fittings have to be the same type as the existing ones, or whether it's permissible to switch to a different type that will be easier to work with. Specifically, does the code permit the integration of plastic

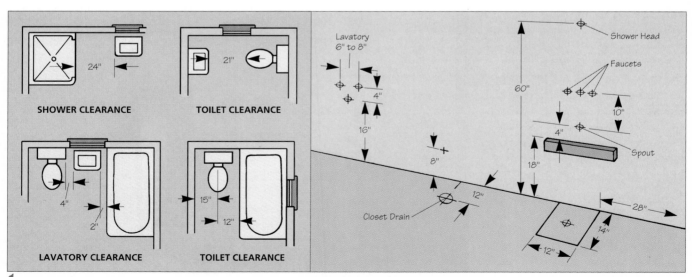

1 *In making the layout, be sure to allow ample clearance between fixtures. Shown is the minimum clearances recommended between bathroom fixtures. When roughing-in plumbing for a new bathroom, first establish the location of each fixture, noting the position of drains and faucets so stub-outs can be cut through walls and floors at the exact spots they are needed.*

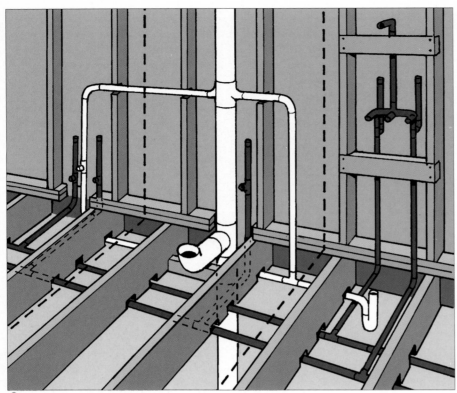

2 *After you've drawn the water and DWV systems in place in detail, as shown, make a list of the parts you'll need. Now you're ready to install the plumbing.*

DWV pipe with an existing cast-iron drain system? Plan to use copper tubing for all water-delivery piping.

2 **Draw the Rough-In Plumbing.** Draw the rough-in procedure to pinpoint the exact spots where water and soil/waste pipes will come into the new bathroom to hook up with a new fixture. Don't take the term "rough-in" literally. The drawing should be a precise, detailed layout of the arrangement that clearly shows where each pipe must go and how you'll get it there.

Running Copper Water Supply Tubing

The hot and cold water supplied to fixtures throughout the house runs through copper tubing (sometimes galvanized-steel pipe in older houses). After tapping into an existing line, simply solder lengths of copper tubing together until the fixture is reached. The lines to each fixture

terminate in a shutoff valve. When adding supply lines, run tubing parallel to floor and/or ceiling joists wherever possible and tuck them into the space between joists.

Soldering Copper Tubing

Determine where you want to tap into an existing water supply pipe. The easiest way to tap into an existing pipe is to find an accessible elbow and replace it with a tee. To do the job, turn off the water supply at the main valve and drain the water out of the system. Use a propane torch to heat the elbow until the solder melts. Wear heavy-duty leather gloves, and use pliers to pull off the elbow once the solder has melted. Clean both pipe ends thoroughly, as well as the inside of the tee. Apply flux, then solder the joint together.

Difficulty Level: 🔩🔩

Tools and Materials

☐ Basic plumbing tools
☐ Small brush
☐ Solder
☐ Flux

1 **Shut Off the Main Water Supply.** Locate the main water shutoff valve and turn it off com-

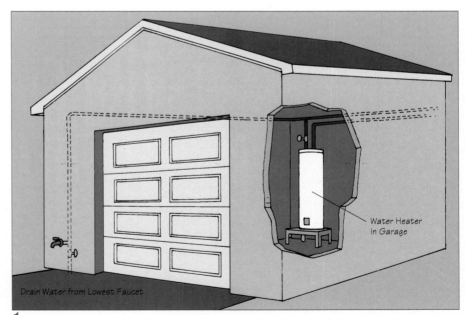

1 *Close the main water shutoff. Open all the faucets, especially the one located closest to the ground (usually an outdoor spigot) to drain the water out of the system.*

pletely. Open both hot and cold water faucets throughout the house to relieve pressure and help drain residual water from the system. Be sure to open a faucet or spigot that's located at the lowest point of the system to assist drainage. Keep the faucets open until all soldering work is complete and shutoff valves have been attached to the pipe stubs in the new bathroom.

2 **Cut the Copper Tubing.** Determine the length of tubing needed and use a tubing cutter to cut it. Lock the cutter onto the tubing and turn the tool one complete revolution. Tighten the cutter's handle about one-half a turn and rotate the tool again. Continue tightening and turning the tool this way until the cut is completed.

3 **Ream Out the Cut End.** Most tubing cutters have a reamer attachment (a pointed blade) to remove burrs from the inside edge of the cut. If you don't have a reamer, lightly file the inside perimeter of the cut with a round file.

4 **Clean the Tubing Ends.** Use emery cloth or a pipe-cleaning wire brush tool to brighten the ends of the tubing and the inside sections of fittings into which it will fit. This step is critically important, as solder will not adhere to a dirty copper surface.

5 **Apply Flux.** Unless the solder you use contains flux, you'll have to apply a film of flux to all copper tubing ends and fittings before slipping the pieces together and soldering them. Use the small brush that came with the flux to apply a light coat to the tubing ends and fitting interiors. Slide the tubing and fittings together.

6 **Solder the Joint.** With the pipe and fitting put together, be certain that the length and angle are correct before applying heat from a propane torch. Heat the fitting (not

the pipe) for about 5 seconds, then touch the pipe/fitting joint with solder. When the fitting and pipe are hot enough, the solder will suddenly

melt and flow into the fitting and surround the outer perimeter of the pipe. When solder starts to drip from the fitting, the job is done.

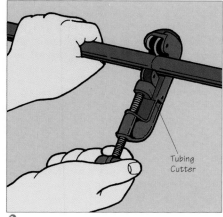

2 *Use a pipe cutter to cut pipe to length. You can use a hacksaw if you don't have a pipe cutter.*

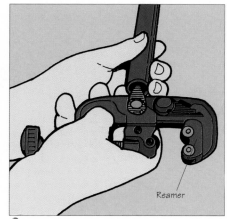

3 *Remove burrs from the inside of the copper pipe with a reamer, which may be a part of the cutter you have.*

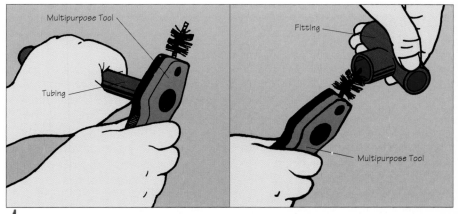

4 *Clean the ends of the pipe with emery cloth or a special wire brush as shown until they shine brightly. You must also clean the insides of fittings.*

5 *Apply a light coating of flux to the ends of the pipe and inside portions of the fittings.*

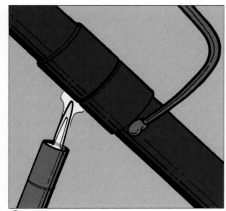

6 *Heat the fitting for 5 seconds until the solder melts and flows into the connection. When solder starts to drip, stop.*

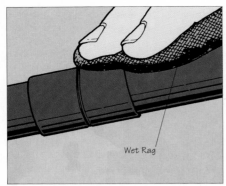

Once you've soldered the connection, use a sopping wet rag to wipe the joint and cool the pipe and fitting.

Cool the Joint. As soon as the soldering is complete, place a sopping wet towel or rag on the connection to cool it and remove excess solder. Be extremely careful, as the pipe and fitting will be hot.

Relocating Existing Supply Lines

Pipes in a basement that supply water to fixtures upstairs are often attached to the underside of the floor joists. If a drywall ceiling is to be installed in the basement, the pipes have to be relocated. The existing pipes may be reused or it may be easier to replace the old pipe with new. You can relocate the lines if there aren't many pipes to move; however, if the house

Fire-Safety Caution

Since the flame from a propane torch can easily ignite wood and other combustibles, play it safe and cover nearby framing members or other flammable material with a cookie sheet or fireproof cloth. These cloths are readily available at plumbing supply stores, hardware stores, and home centers. Look for them in the plumbing department.

has hot water or steam radiators, there will be too many pipes to move. Consider a suspended ceiling instead.

Difficulty Level:

Tools and Materials

☐ Basic plumbing tools
☐ Marker
☐ Saber saw (or handsaw)
☐ Metal plates for joists

Mark the Runs. Run a carpenter's pencil or a marker along the sides of the existing horizontal pipes to mark the underside of the joists for notches. The pipes most commonly encountered are ½-inch, ¾-inch, and 1-inch (inside diameters). The outside of a copper water pipe

is about ⅛ inch larger than its internal dimension, so notches should range from about ⅝ inch wide to 1⅛ inch wide.

Measure for Cuts. Before removing the pipes, figure out how much they have to be raised. To minimize the depth of notches in the joists, the bottom of the pipes can sit flush with the bottom of the joists. Measure from the bottom of a pipe to the underside of a joist. Add at least ¼ inch to this dimension to allow for pipe fittings. The total is the amount that needs to be cut off each riser in order to raise the pipes.

Cut the Risers. Use a propane torch to liquefy the solder in the existing fittings so you can discon-

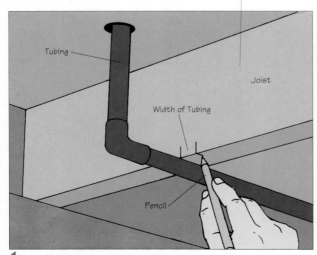

Mark the location of the notch on the underside of each joist.

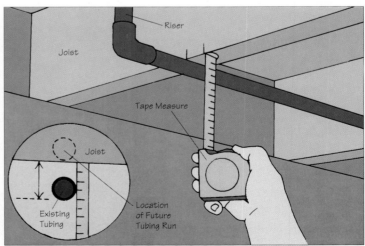

Measure from the bottom of the pipe to the underside of a joist. Add ¼ in. to get the cutoff distance.

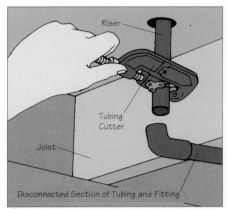

3 *Use a tubing cutter to cut each riser, then prepare to re-solder the pipes and fittings.*

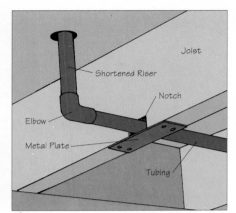

4 *Fit the pipe runs into the notches and re-solder all the fittings. Nail a metal plate over each notch.*

nect them. Then use a tubing cutter or, if space does not permit, a mini-cutter or hacksaw to cut the risers. Remove a piece equal to the distance the pipes are to be raised. Then use a handsaw or saber saw to cut notches in the joists. Use the marks from Step 1 as a guide.

4 **Reconnect the Pipes.** To promote a good solder joint, clean out the fittings and pipe ends on the existing pipe runs. Then lift the runs into place and solder them together. Nail a metal plate over the notches to protect the pipes from possible drywall nail or screw punctures.

Drainpipe Systems

In most cases you should be able to install a new plumbing drain system using PVC (polyvinyl chloride) plastic pipe. Check with your local building department. You connect PVC pipe to fittings with special PVC cement. If the existing DWV system consists of cast-iron pipe, you'll have to use a "no-hub" neoprene fitting to join a piece of PVC pipe to an existing

cast-iron pipe as a means to begin the new drain line.

Toilets generally require a minimum 3-inch drain line; bathtubs and showers require a minimum 2-inch drainpipe; and bathroom lavatories require a minimum 1½-inch drain. Each line must be vented to the atmosphere through a vent pipe (usually the main stack). The National Standard Plumbing Code allows wet venting (a drain line only to the main DWV stack, which includes a main vent line) for a lavatory when it's within 30 inches of the main vent stack and bathtub/showers when they're within 42 inches of the main stack. Units located farther away from the main stack must have an additional vent pipe coupled to the drain line that runs up to and then connects with the main vent stack.

Connecting Drain Lines

No-Hub Fitting. Determine where you want to tap into the cast-iron DWV system and either dismantle an existing elbow or cut the pipe to make way for a tee or elbow. Slip the no-hub fitting onto the cast-iron

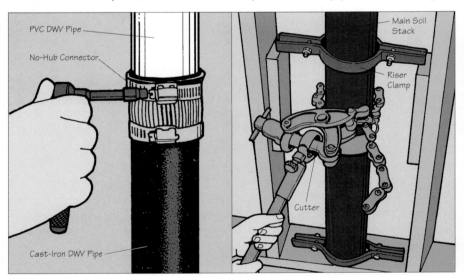

Drainpipe Systems. *Use a no-hub neoprene connector to join new PVC plastic pipe to existing cast-iron pipe when extending a drain line to a new bathroom. A snap or chain cutter makes cutting through cast-iron pipe a fairly quick job. A riser clamp supports the weight of the cast-iron pipe.*

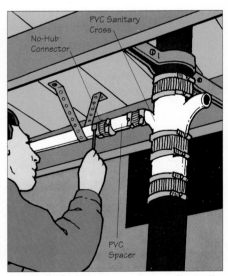

No-Hub Fitting. *Generally, you'll install a T- or four-way fitting onto an existing drainpipe with a no-hub connector to add an inlet for a new bathroom.*

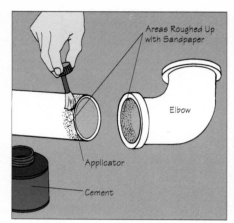

PVC Cement. *For best adhesion, scuff up the outer ends of the pipe and the inner portions of the fittings with sandpaper before applying the cement.*

piece, then insert the PVC pipe section. Apply the stainless-steel band and screw clamps over the neoprene sleeve and tighten the screw clamps.

PVC Cement. All other PVC connections will be made with cement. Clean and roughen the ends of PVC pipe and interior sections of PVC fittings with sandpaper to remove the sheen from the plastic and enable the cement to make a solid connection. Be sure the connection will fit correctly before applying cement. PVC cement dries quickly; you'll have only a couple of seconds to adjust the pipe and fitting before it sets and the joint is solid.

Securing Pipes

All plumbing pipes must be secured to joists, studs, or other framing members to keep them in place, prevent bowing, and keep unnecessary stress from weakening joints. A number of clamps, hangers, and brackets are available for every pipe size. Some fit over pipe and are nailed to wood framing members. Others clamp onto pipe and are then supported by hangers. It's imperative that water pipes

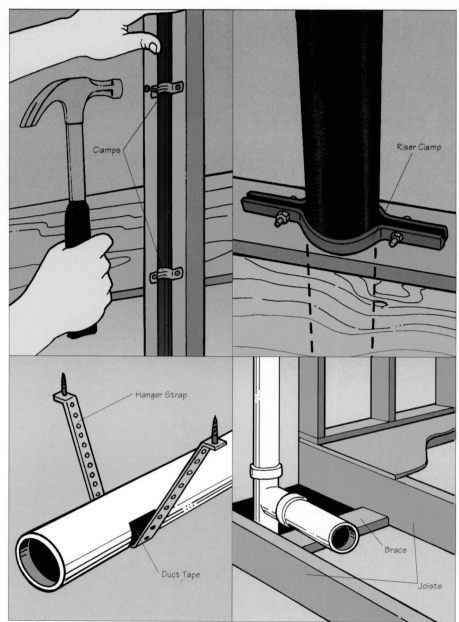

Securing Pipes. *Clamps and hangers are used to support pipes. Support water pipes every 48 inches or less and at every turn (A). Support vertical runs of DWV pipe with riser clamps (B) and horizontal runs with hanger straps (C). Place duct tape or electrician's tape between the pipe and hanger. Horizontal overhead lengths of DWV pipe that run parallel to joists can be supported with wood braces (D).*

be secured to framing at least every 32 inches of run. This helps to reduce vibration in the pipes and holds them secure when water is suddenly shut off and creates a water-hammer effect. The same

is true for drainpipes, although they're under no internal pressure. If need be, especially for a large toilet drainpipe, nail a 1x4 brace between joists to serve as an anchor point for the pipe's clamp, hanger, or bracket.

Finished Walls and Ceilings

9

After completing all of the framing, electrical, and plumbing work in your conversion, it's time to apply the finishing touches that will determine the character and comfort of a room. For the most part, this will entail installing drywall over wood framing for ceilings and walls. It's important that you finish these materials so they match or blend with the rest of the house. Remember, the objective is to make the conversion appear as if it were part of the house's original construction—not a conversion. This concern is paramount whenever you remodel any living space inside a home.

Posts and Beams

All framing work must be completed before you can finish the walls. This preliminary work includes furring for concrete foundation walls, framing around heat ducts, and preparatory work around posts and beams that may be situated in the middle of a new conversion. Posts and beams are part of the structural system that holds up a house and must never be altered, moved, or eliminated without the guidance of a structural engineer. Those structural members are often in the way when it comes to remodeling plans—sometimes in attics that feature structural ridge beams, in garages that support living space above, and almost always in basements.

Working with Posts

Posts typically provide intermediate support for beams. In most cases, they're found in garages and basements on top of concrete slabs that have been thickened to form a footing that distributes the structural loads. The top of each post is toenailed or bolted into place to prevent lateral movement. Posts in older houses are usually made of solid wood; those in newer houses are usually lally columns. A lally column is essentially a steel tube that can be adjusted to various heights. The columns range from 3 to 5½ inches in diameter and may be filled with concrete. They're sometimes secured to a wood beam with nails that run up through the top flange.

"Buried" Posts. If a post isn't ideally placed in relation to remodeling plans, try to revise the plans rather than remove the post. Moving a post is an option of last resort. One or more posts may be concealed by "burying" them in a wall that separates two rooms. If the post is unusually big in diameter, you can

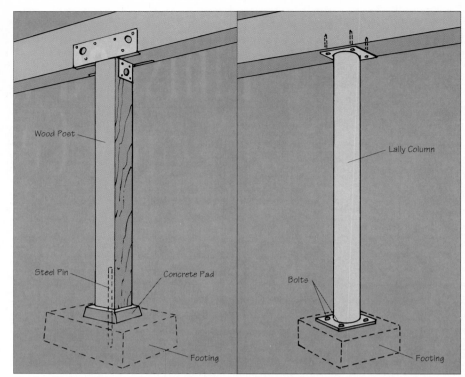

Posts. *The posts located in the basement rest atop a footing of some type. A lally column is a steel post that ranges from 3 to 5½ in. in diameter. It's sometimes secured to a wood beam with nails that run upward through the top flange.*

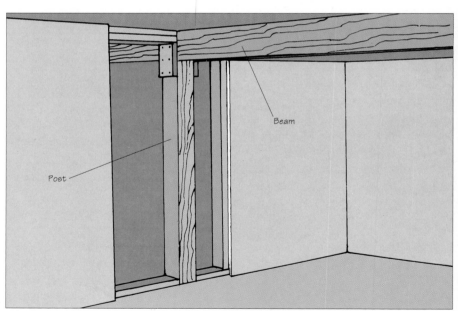

Buried Posts. *A wood or steel post can be covered with partition walls. Unusually large posts might require a wall framed with 2x6 lumber.*

frame the wall with 2x6 lumber rather than 2x4s.

Concealed Posts. If it's not possible to bury the post, you can disguise it. You might nail plywood paneling or

drywall to a wood post and treat the edges just as you would walls finished with the same materials. Or simply sand the post smooth, round over or chamfer the edges with a router, and paint it. If the

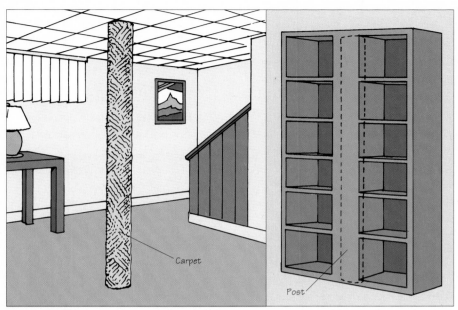

Concealed Posts. *You can apply carpeting to a post to conceal it. Another option is to build a set of open shelves around the post.*

post is metal, there are several ways to conceal it. One possibility is to apply carpet to it. First test-fit the carpet to make sure it's the right size. Then spray or brush both the column and the carpet backing with contact adhesive. When the adhesive becomes tacky, wrap the carpet around the column. Another option is to build a shelving unit around the post.

Framing around a Post

One of the best ways to conceal a lally column is to frame around the column using two-by lumber. The frame provides a base for other finishes.

Difficulty Level:

Tools and Materials

☐ Basic carpentry tools
☐ 2x3s or 2x4s
☐ Masonry nails
☐ Common nails, 12d
☐ Drywall or paneling
☐ Drywall nails or screws
☐ Corner trim or corner bead
☐ Metal-cutting snips

1 **Lay Out the Frame.** The outside dimensions of the box can be any size as long as the inside

dimensions are large enough to accommodate the post. It's usually best to minimize the overall size of the box, however, to keep it from overpowering the room. Use a framing square to lay out the inside dimensions of the plates.

2 **Install the Frame.** Use 2x3s or 2x4s for the framing lumber. Assemble two opposite "walls" of the frame to fit between the beam and the floor. Using the layout lines as a position guide, tip the walls into place. Then use a level to make sure the frames are plumb. Nail the plates to the floor (with masonry nails if the floor is concrete) and to the underside of the joists above. Cut blocks to fit between the frames at the top and bottom. Toenail the blocks to the plates. If the frame walls are at all bowed, you can add blocks halfway up the walls to straighten them.

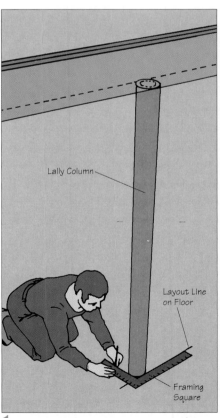

1 *Use a framing square to lay out the locations of the plates. Align the tool so that the layout is perfectly square.*

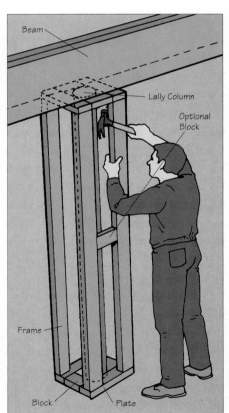

2 *Assemble the two side "walls" of the framing. Use blocking to fill in the spaces between the walls.*

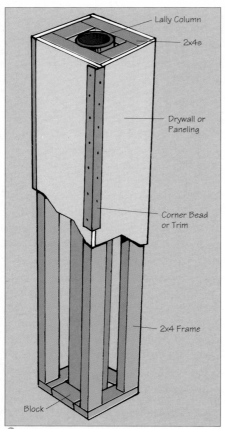

Cover the frame with drywall or paneling.

Apply the Finish Surface. Once the framing is secure, apply drywall or wood paneling with nails or screws. With wood paneling, miter the corners or cover them with corner trim. Use standard corner bead when installing drywall.

Boxing around a Post

Another way to conceal a lally column is to create a wood box to surround it. This method consumes less space than the frame just described. It can only be used, however, if you've installed a plywood subfloor. The box can't be attached to a concrete floor. Use one-by stock to build the box; pine or a hardwood such as oak is appropriate. Pine can be painted or stained, while hardwood can be stained or left natural and coated with a clear varnish or sealer.

Difficulty Level:

Tools and Materials

☐ Basic carpentry tools
☐ Framing square
☐ One-by lumber
☐ Table saw (or circular saw)
☐ Wood glue
☐ Finishing nails, 6d
☐ Sandpaper
☐ Wood putty
☐ Paint or stain
☐ Brush

1 Lay Out the Box. Use a framing square to lay out the inside perimeter of the box. You can then draw the outside perimeter ¾ inch outside of the first line, providing the exact outside dimensions of the box.

2 Cut the Sides. Measure the distance between the floor and ceiling, then subtract ¼ inch from the measurement to provide a fitting allowance. Cut four pieces of stock to length. Use a table saw or circular saw to miter each edge at a 45-degree angle. Test-fit the assembly around the post.

2 *Use a table saw to miter the sides. For safety, use a blade guard (not shown here for clarity) and make sure the blade is facing away from the fence.*

3 Assemble and Install the Box. Spread a thin film of wood glue on the edges and use 6d finishing nails to nail three sides of the box together. Then slip the three sides over the post and nail the fourth side into place. Toenail the box to the floor and to the beam above.

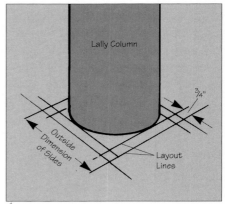

1 *Use a framing square to draw a full-scale layout on the floor.*

3 *Glue and nail three sides of the cover together, then slip the assembly over the post and nail on the fourth side. Toenail the box to the floor and to the beam.*

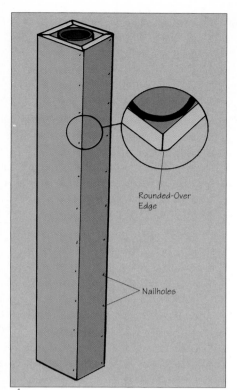

4 *Fill the nailholes with wood putty, then use sandpaper to smooth them and to round over the corners. Stain or paint the box to match the walls.*

4 **Finish the Box.** Use sandpaper to round over the edges of the box. A post is typically located in the middle of a room, and rounding the edges minimizes impact damage, both to the box and to those who may bump into it accidentally. Use wood putty to fill all nailholes, then sand them smooth. Paint or stain the box.

Working with Beams

A beam provides intermediate support for floor joists. As with posts, beams can't easily be removed or relocated, but they don't obstruct floor plans as much as posts do. Remember that building codes call for at least 84 inches of headroom beneath a beam.

In newer homes, steel or glue-laminated wood is typically used for beams that span more than 8 feet. A solid-wood beam is sufficient for smaller spans. Flitch beams, made up of a sandwich of wood and metal, combine the strength of steel with the look and lighter weight of wood.

Beam Connections. The connection between a post and a beam must be rigid. Depending on the kind of post and the kind of column, there are several ways to achieve a rigid connection. A metal saddle is sometimes installed around the beam. This kind of saddle has flanges that are nailed to wooden posts. Another kind of saddle goes around the beam and is welded to a metal post. Two beams that meet over a single post must be connected to one another as well as to the post, usually with a metal strap.

If you plan to apply drywall or paneling to the beam, these connections may get in the way. Never remove a connection without replacing it with something of equal strength.

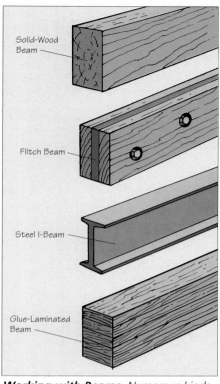

Working with Beams. *Numerous kinds of beams can be used to support joists in a basement. Shown here are several of the most common.*

Concealing a Beam

The task of concealing a wood beam, like that of concealing a post, isn't difficult. A steel beam, on the other hand, isn't easy to conceal because it's difficult to fasten material to it. To get around this problem, you can secure paneling or drywall to wood framework that's nailed to the underside of the ceiling joists.

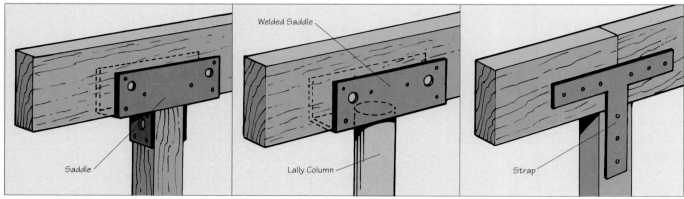

Beam Connections. *There are a variety of ways to secure a beam to a post. Although these connections sometimes get in the way when drywall is being installed, they must never be removed without replacing them with something equally strong.*

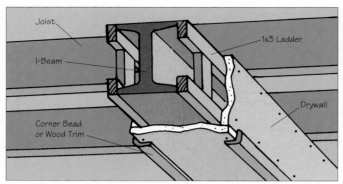

Drywall on Steel. *Build "ladders" to support the drywall that you'll use to cover a steel I-beam.*

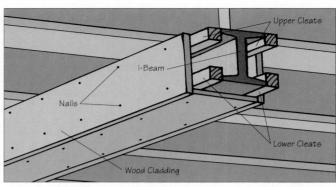

Wood on Steel. *Use a simple cleat system to support solid-wood cladding around a steel I-beam.*

Drywall on Steel. First build two wood "ladders" made of 1x3s. Place the ladders against the beam and toenail them to the joists. Attach drywall on all three sides. Before finishing the walls, cover the drywall joints with trim or corner bead—a metal strip made at a 90-degree angle designed to be nailed to outside drywall corners.

Wood on Steel. To make the job of covering the beam easier, use ⅜-inch or thicker paneling, ½-inch plywood, or ¾-inch solid wood. Attach cleats to the joists against both sides of the beam. Then use glue and finishing nails to attach cleats along one inside edge of the side panels, allowing for the thickness of the bottom panel. Nail through the panels into the cleats. Glue and nail the side panels to the upper cleats. Then glue and nail the bottom panel to the lower cleats.

Concealing Ducts

A large rectangular sheet-metal duct called a trunk often leads from a furnace to the farthest points of a house. Trunks are commonly found along attic floors and basement ceilings. Smaller ducts branch off the trunk and distribute warm air to each room served. The ducts in a central air-conditioning system may have a similar layout. If the ducts obstruct headroom, it may be possible to move them, but this definitely is a job for a heating-and-cooling contractor. In most cases, it's easier and less expensive to leave the ducts in place. With an informal decor, you can just paint the ductwork to match the ceiling color. It may also be possible to enclose them within the confines of a suspended ceiling. If not, you can box the ducts within wood framework covered with drywall or paneling.

Concealing Soil Pipes

The soil pipe, the main drainpipe of the plumbing system, conducts water and waste away from the house. Typically, it's the largest pipe in the house and may be plastic or cast iron. If possible, enclose the pipe within a box or soffit. It's a good idea to wrap the pipe in insulation before you box it in, especially if it's plastic. The insulation reduces the sound of rushing water. Be sure to take measurements in several places along the length of the soil pipe before making the box because the pipe slopes at least ¼ inch per foot for proper drainage. If the pipe's clean-out plug will be covered up by the concealment process, include a door for access to the plug.

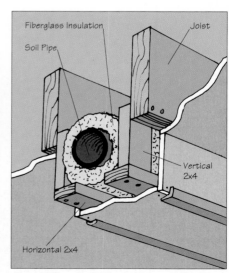

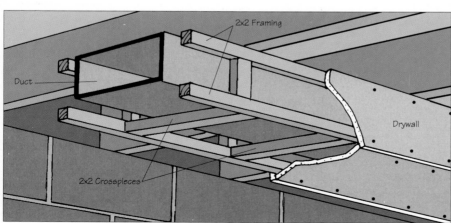

Concealing Ducts. *Conceal heating and air-conditioning ducts as if they were beams. You can enclose ducts that run alongside beams in the same box.*

Concealing Soil Pipes. *You can also box-in a soil pipe as if it were a beam or duct.*

Finishing Walls and Ceilings

Just about any wall or ceiling surface can be used in an attic, basement, or garage conversion, including solid-wood strip paneling, sheet paneling (plywood and others), drywall, acoustic panels, and even bare concrete block.

Painting Masonry Walls

You won't find a masonry wall, except for a large chimney, in any ordinary attic, but some garages and most basements are framed with them. You can go to great lengths to try to make a basement or garage conversion seem less like the utilitarian space it is, but maybe your goal is simply to brighten up the space without devoting too much time to the project. If maximum impact and minimum expense and effort are what you're after, consider painting the masonry walls. Poured-concrete and concrete-block walls can be successfully painted, but the paint can't cover defects. Cracks, flaws, and poor mortar joints show through just as clearly after painting as before. In fact, they may stand out even more when the walls become a consistent color. Try to make the wall as defect-free as possible by filling voids with hydraulic cement, then scraping and wire brushing the surface.

For paint to adhere properly it must be applied to a surface that's free of dirt, dust, and grease. In a basement or garage, it's especially important that the surfaces be dry. If the wall is susceptible to moisture problems, the problems must be corrected first or the paint will flake off the wall (see "Sealing a Masonry Wall," page 56). The care you invest in cleaning, scraping, and

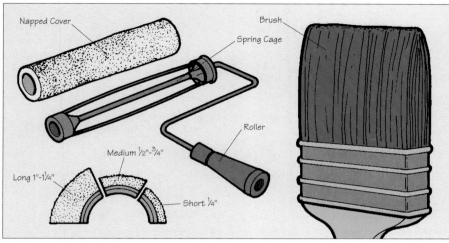

Brushes and Rollers. *Masonry surfaces, particularly concrete block, are fairly rough. To paint these surfaces quickly, equip a roller frame with a long-napped roller cover. You'll need a brush for touching up corners and edges.*

patching the walls makes the effort of painting worthwhile.

Types of Paints. Standard latex paints are water-based, easy to clean up, and quick to dry. Instead of water, oil-based paints use natural or synthetic oil as the vehicle for carrying the pigments and binders. Oil-based products take longer to dry and require solvents for thinning paint and cleaning tools. These paints also call for plenty of ventilation to dispel fumes. Basements and below-grade garages aren't easy to ventilate, so it's better to use latex paint. No matter what kind of paint you use, you must clean and prime the masonry surfaces before you can begin painting.

Brushes and Rollers. You can paint concrete foundation basement and garage walls entirely with a brush, but you'll do the job faster with a roller. A roller consists of a frame and a cover. Covers vary in thickness and nap composition. A short nap, about ¼ inch thick, applies a thin, smooth layer of paint and is suitable for the smooth surface of a poured-concrete wall. A longer nap, about 1 inch thick, deposits a large amount of paint and is better for porous or irregular surfaces, such as concrete block. Other necessary items include

a roller pan for loading the roller with paint and a 3- or 4-inch paintbrush for painting into corners and around details. A natural-bristle brush is best for oil-based paint; use a synthetic-bristle brush for latex paint. The water in latex paint ruins natural bristles.

Types of Drywall

Drywall, also known as wallboard, plasterboard, and gypsum board, or by the tradename Sheetrock, is the most commonly used material because it's versatile and relatively inexpensive. Some codes require that drywall be installed beneath other wall surfaces to provide a measure of fire safety. Regular drywall has a gray kraft-paper backing. The front is covered with smooth off-white paper that takes paint readily, although you should coat it with a primer before applying paint. The long edges of each 4x8-foot sheet are tapered slightly to accept tape and joint compound. Standard drywall comes in several thicknesses: ½ inch is usually appropriate for most conversions, but ⅝-inch-thick drywall better resists bowing between ceiling joists spaced on 24-inch centers. Some kinds of drywall have special purposes. Water-resistant drywall, usually faced with blue or green paper, is made for use in areas of high moisture, such as bathrooms. Fire-resistant drywall, called Type-X, is required by some local building codes around furnace enclosures and other combustion appliances and on ceilings or walls that separate garages or workshops from living spaces.

Tips for Nailing Drywall

• Use 1⅜-inch-long ring-shank drywall nails to install ½-inch drywall.

• Space nails 5 to 7 inches apart on ceilings, 8 inches apart on walls. Don't nail closer than ⅜ inch or further than ½ inch from the edge of a sheet.

• Set the head of each nail slightly below the surface of the drywall. This technique is called "dimpling," and it allows the nailhead to be concealed with joint compound later on. A standard hammer can't make a proper dimple because it's likely to damage the face paper. Instead, use a special drywall hammer, which has a slightly convex face. In place of claws, these hammers have a sharpened edge that you can use to trim drywall quickly.

• If you make the mistake of driving a nail too hard and the face paper breaks, the nail won't hold. Drive a second nail nearby to ensure proper holding.

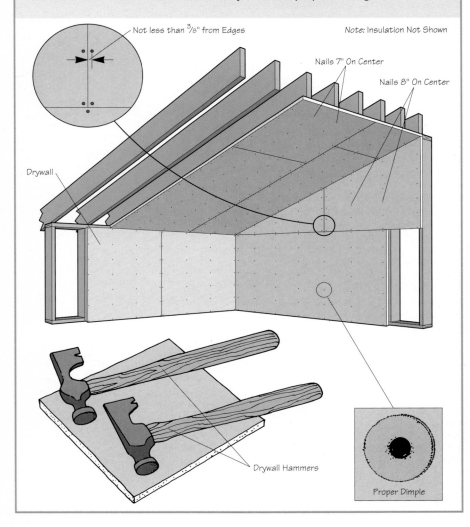

Not less than ⅜" from Edges

Note: Insulation Not Shown

Nails 7" On Center

Nails 8" On Center

Drywall

Drywall Hammers

Proper Dimple

Installing Drywall

Difficulty Level: 𝕋 𝕋

Tools and Materials

☐ Basic carpentry tools
☐ Keyhole saw
☐ Drywall (½- or ⅝-inch)
☐ Lipstick
☐ Aluminum 48-inch drywall T-square (or straightedge)
☐ Electric drill with Phillips bit for screws
☐ Drywall nails or screws long enough to penetrate at least ¾ inch into the framing

1 Estimate Your Needs. Calculate the square footage of the ceiling and each wall. Then add these figures together to get a total for the entire room. Add 10 percent to this figure to account for waste, then divide by 32, or the number of square feet in one 4x8-foot drywall sheet. The result is the approximate number of drywall panels needed for the job.

2 Cut the Drywall. Place one sheet on a flat work surface and use a utility knife guided by a straightedge to score the face paper. Drywall-cutting T-squares and straightedge guides are readily

1 *Measure wall and ceiling surfaces to determine the approximate square footage. Measurements need not be precise. Round to the nearest 6 in.*

2 Score a straight line on the face of the drywall sheet and snap it along that line. Keep your fingers well away from the cut line when using a sharp knife.

3 With at least two people supporting a sheet, press the drywall against the ceiling joists and nail it from the center of the panel toward the edges.

available at home centers and lumberyards. Shift the drywall so the score line overhangs the work surface, then snap it along the line. Slice through the paper backing to remove the piece.

3 Install the Ceiling Drywall.

Install drywall on ceilings first, then on walls. A 4x8-foot sheet of ½- or ⅝-inch drywall is heavy and awkward to handle, so you'll need assistance when it comes time to install the material on the ceiling. (Drywall lifts make this job easier, and possible for one person. The lifts are available at most tool-rental yards.) Cut a sheet to size, then lift it into place and hold it firmly against

the rafters or joists. Quickly nail or screw the sheet in several locations at the center to hold it in place. With a helper or two (or the drywall lift) supporting the sheet, use a chalk-line to mark the positions of rafters or joists. You can also make a simple T-brace from 2x4s to help hold the sheet in place. After you've marked the sheet, complete the nailing or installation of drywall screws.

4 Install the Drywall on the Walls.

Install the drywall vertically on partition walls so that the tapered edges of each sheet fall over a stud. You can also apply the sheets horizontally to minimize the number of seams between them.

4 Use a lever to lift the sheets into place or rest the sheet on a small scrap of wood.

The Screw Alternative

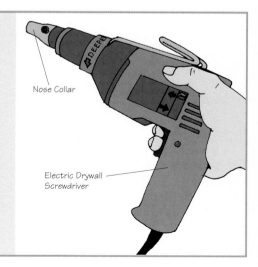

Drywall nails are the original way of attaching drywall. A newer method is to use drywall screws. You can drive these screws with any variable-speed electric drill equipped with a Phillips bit. Or you can buy or rent an electric drywall screwdriver, sometimes called a screw shooter. This tool is a high-speed electric screwdriver with an adjustable clutch and nose collar. The clutch stops driving the screw when it's just below the surface of the drywall, but before it breaks the paper, creating a dimple that's smaller and easier to fill with joint compound than the dimple produced by a hammer. Use drywall screws that are at least ¾ inch longer than the thickness of the drywall. If you're a novice drywall installer, use a drywall screwdriver rather than an electric drill to ensure that screws are driven to a consistent depth each time and to avoid going too deep, causing damage to the drywall.

5 To accommodate obstructions like electrical boxes, apply lipstick to the box, press the drywall against the box to mark its location, and cut the hole.

If you've got to lift the panel slightly in order to maneuver it into position, use scraps of wood to make a foot-operated lifting lever. Butt the top edge of the lower panel firmly but not forcibly against the ceiling panel, which is already in place.

5 **Mark for Cut-Outs.** Cut holes in the drywall to accommodate electrical boxes. Use lipstick to mark the outside edges of each box, then cut a sheet of drywall to the proper size and push the sheet firmly into position. The lipstick marks where to make the hole. Use a keyhole saw or special drywall saw to cut the opening. Stagger all wall and ceiling joints.

Finishing Drywall

After you've installed the drywall you must conceal the seams and nail or screw dimples, as well as imperfections such as accidental gouges, in a multistep process called "finishing." You must do this work meticulously because even the smallest dents and ridges will show through paint and wallpaper.

Tools and Materials. A thick paste-like material called joint compound is the main ingredient for finishing

drywall. You spread several layers of compound over imperfections and sand each layer smooth after it dries. Embed inexpensive paper tape, which is sold by the roll, in the compound to reinforce seams and prevent cracks from appearing at these locations. You can use conventional 100-grit sandpaper to smooth out dried joint compound, but open-weave silicon-carbide paper is better because it doesn't become clogged with sanding dust as standard papers do—and it lasts longer. Better yet, if you can find it, is drywall sanding mesh, which doesn't clog at all.

You'll need several taping knives of different sizes to spread the compound and smooth wet seams. Each knife has a thin flexible steel blade. Knife size is partly a matter of personal preference, but a 4- or 5-inch broad knife is good for applying the first coat and for filling nail or screw dimples, and 10- and 14-inch knives are good for finishing joints.

CAUTION: Wear goggles and a dust mask or respirator when sanding.

Difficulty Level: 🔩🔩

Tools and Materials

☐ Joint compound
☐ Mud pan (or old loaf pan)
☐ Drywall knives (4-, 6-, 10-, 14-inch)
☐ Joint tape
☐ Metal-cutting snips
☐ Metal corner bead
☐ Sandpaper (100-grit) or drywall mesh
☐ Sanding pole
☐ Goggles
☐ Dust mask

1 **Fill Flat Joints.** Use a wide knife to drop a gob of joint compound into the mud pan or an old loaf pan from the kitchen. Use the smallest knife to force the compound into the tapered drywall joints until they're filled and level. At butt joints, where the nontapered ends of two panels

meet, fill the crack and create a slight hump, which will be flattened later.

2 **Embed the Tape.** Cut a piece of joint tape to length and center one end of it over the joint. Embed the tape by using the smallest knife to smooth it into the compound. Spread a 1/8-inch-thick layer of compound over the tape. Hold the knife at a 45-degree angle and go back over the joint to scrape away excess compound.

1 Use a 4-in. knife to spread joint compound over a seam. The compound will fill the tapered area between two drywall sheets.

2 Cut paper tape to the length of the seam, center the tape over the seam, and use a small knife to smooth the tape into place.

3 Finish Inside Corners. Start by filling both sides of inside corner joints with compound. Fold a length of paper tape along its precreased centerline and apply it to the joint. Remove excess compound as in Step 2. If you find the inside corners difficult to finish, use an angled taping knife, called an inside corner taping tool, to make the task easier.

4 Finish Outside Corners. If there are any outside corners, use metal-cutting snips to cut a length of metal corner bead to the height of the wall corner. Angle the cut ends inward slightly to ensure a better fit, and use drywall nails rather than screws to attach the bead to the wall corner. Using the corner edge of the bead to guide the knife, fill the bead with joint compound.

5 Finish Odd-Angled Corners. Corners other than 90 degrees, such as the intersection of the knee-wall and the ceiling in an attic, are difficult to finish using standard paper tape. A solution to this problem is to use multiflex tape, paper tape that's reinforced with two metal ribbons. The metal side faces the wall and stiffens the tape just enough so that it can form a consistent obtuse angle. Multiflex tape is also particularly good for the outside corners of skylight wells where the corners may be other than 90 degrees.

6 Fill Nail or Screw Dimples. Use the smallest knife to fill all nail or screw dimples and other minor imperfections with compound. You don't need tape for this procedure.

7 Apply Finish Coats. After the first coat of joint compound is dry, inspect the seams and remove all ridges that might interfere with the smoothness of subsequent applications. Scrape off ridges with the 4-inch knife. Then use a 10-inch knife to apply a thin second coat of compound to the joints and to go over the dimples again. After the second coat dries, scrape it and use a 14-inch knife to spread the third and final coat. Spot dimples only if they're not filled completely.

8 Sand the Compound. After 24 hours, or when the compound is completely dry, sand all joints and dimples until they're smooth. Fold the sandpaper or drywall mesh into quarters and go over the compound lightly. Be

3 *Use a special corner knife to make it easier to finish inside corners.*

4 *On outside corners, pull the knife along the edge of the corner bead.*

5 *Metal ribbons on multiflex tape make it easier to use on odd angles.*

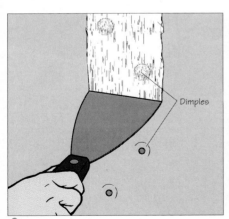

6 *Cover nail dimples and other imperfections with at least two coats of joint compound.*

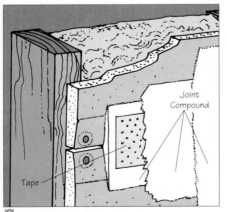

7 *Three layers of compound spread in successively wider swaths make a finished joint.*

8 *Sand joints, using a pole sander as necessary. Shine a bright light across the walls to check your work.*

careful not to sand through the face paper. After sanding, shine a bright light across the walls and ceilings to detect any imperfections in the finished surface. If a section of tape has bubbled because it didn't adhere properly, cut the bad section out and apply another thin coat of joint compound to the area to smooth it.

Use a universal pole sander to make the work easier. This tool has a pad with clamps to hold sandpaper or drywall mesh. The pad is swivel-mounted to a pole. A sanding pole is particularly handy for reaching ceilings, but you may find yourself using it almost everywhere because it extends your sanding stroke. Brush or vacuum away all traces of sanding dust before texturing, painting, or papering the walls. Wear a dust mask or respirator because the dust is plentiful and extremely fine.

Painting Drywall

Cover new drywall with a primer before applying paint. Drywall manufacturers recommend a product called P.V.A. Drywall Primer, but you should ensure that the primer you apply is compatible with the paint you'll use. Purchase the same brand of primer and paint to be sure they're compatible. Primers set up wall surfaces to accept paint more readily, and paint adheres best when applied over a primer. For enamel paint, the manufacturer may recommend that an undercoater product be applied before paint.

Wallpapering Drywall

A huge assortment of wallpaper patterns and styles is readily available at home centers and specialty outlets that carry wallpaper, paint, and floor coverings. You can buy wallpaper prepasted or unpasted, whichever type you prefer to install. Like primer for paint, sizing is used beneath wallpaper. You must first clean the walls,

then apply sizing to prepare the surface to accept wallpaper adhesive. Sizing also strengthens the bond between the glue and the surfaces to be bonded. You'll need a sharp razor knife to cut wallpaper cleanly. Wallpaper tends to dull razor knives quickly, so have plenty of spare blades on hand.

Suspended Ceilings

Deciding what kind of ceiling to install in a new basement or garage room isn't merely a matter of appearance. More importantly, it's a matter of how much headroom there is and how many ducts, pipes, and wires crisscross the underside of the joists. Often, the best way to solve the latter issue is to install a suspended ceiling system if a lack of headroom doesn't preclude it.

A suspended ceiling, sometimes called a dropped, or exposed-grid,

ceiling, is a grid-like framework of metal channels that hangs beneath the joists on short lengths of wire. The metal channels support lightweight acoustical panels that form the finished surface of the ceiling. The beauty of this system is that it conceals obstructions attached to the underside of the joists, yet allows easy access for fixing pipes or adding wiring later on. Another advantage is that you level the ceiling as you install it; the existing joists need not be level or even straight. A suspended ceiling system also makes the job of installing ceiling lights easier: Simply remove an acoustical panel and replace it with a special drop-in fluorescent fixture.

Anatomy of a Suspended Ceiling

There are five key parts of a suspended ceiling system:

▶ Main runners are the primary support members and are arranged in parallel rows that run the length of the room. Main runners come in a variety of finishes and shapes.

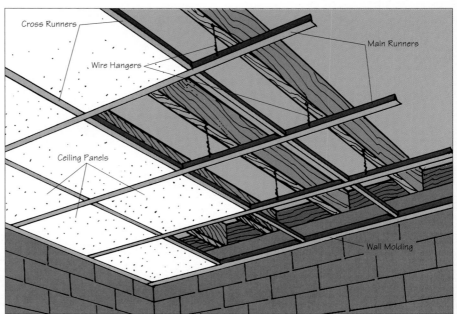

Anatomy of a Suspended Ceiling. *A system of runners, wall moldings, hangers, and panels makes up a suspended ceiling that hangs just below the ceiling joists to cover pipes, ductwork, and other obstructions.*

▶ Cross runners are lighter-gauge supports that fit between the main runners at right angles.

▶ Hangers are lengths of lightweight (usually 18-gauge) wire. You hook one end into holes in the main run-

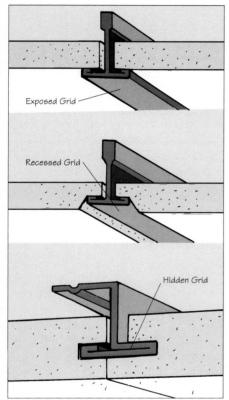

ners and attach the other end to the ceiling joists.

▶ Acoustical ceiling panels fit into the grid that's created by the runners. The panels can be square or rectangular and come in a variety of sizes and patterns.

▶ Wall molding is a metal channel that you attach to the walls. The molding supports ceiling panels around the perimeter of the room.

Types of Runners. There are three styles of runners, each of which gives the ceiling a different look. The runners can be completely exposed beneath the panels, recessed into lips on the panel edges, or hidden in slots in the panel edges.

hacksaw, plumb bob, and utility knife. There are also special tools needed for the job.

Aviation Snips. This tool easily cuts the light-gauge metals used to support acoustical ceilings. The snips are designed to bring maximum leverage to bear on the workpiece and have a spring action that opens the tool after a cut has been made. Snips are available with straight, right-hand, or left-hand handles.

Water Level. The premise behind this tool is simple: Water always seeks its own level, so water contained in clear plastic tubing can be used to locate points around the room that are exactly at the same level. Though professional ceiling installers use a laser level to do the same thing, a simple water level is inexpensive and foolproof and can be purchased at hardware stores and home centers.

Types of Runners. Runners are the primary support system for the ceiling. The look of the ceiling changes according to the type of runners used.

Tools

Many of the tools needed for this project are basic: A hammer, chalk-line, combination square, level,

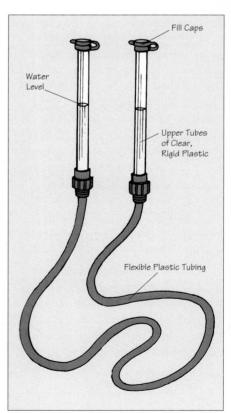

Tips for Using Snips

■ Wear work gloves when using snips. The cut edges and corners of metal ceiling track are sharp.

■ Cut near the throat of the tool rather than near the blade tips to improve control over the cut.

■ Don't tilt the snips as you cut. Doing so twists the edges of the metal and results in rough edges.

■ Don't snap the snips closed at the end of each cut. Doing so bends the metal slightly. Instead, take a good "bite" out of the metal and stop the cut short of the tips. Then take another bite to continue the cut.

Aviation Snips. The leverage provided by the hinged jaws of the snips makes it easy to cut metal ceiling runners.

Water Level. This tool is used to locate points around the room that are exactly at the same level.

Installing a Suspended Ceiling

Difficulty Level: T T

Tools and Materials

☐ Ceiling panels
☐ Runners and wall molding as needed
☐ Installation hardware
☐ Water level
☐ Pencil
☐ Measuring tape
☐ Chalkline
☐ Aviation snips
☐ Hammer
☐ Box nails, 6d
☐ Layout strings
☐ Framing square
☐ Utility knife
☐ Straightedge
☐ Lightweight gloves
☐ U-shaped channel molding
☐ Pop riveter and rivets

1 Plan the Job. Panels are available in 24x24-inch and 24x48-inch sizes. The latter works better if you'll use fluorescent lighting. The panels fit the standard fluorescent tube length. Smaller panels require more cross runners, so the job is more time-consuming. Wall molding and main runners are sold in different lengths up to 12 feet and can be overlapped to reach greater distances. Cross runners are 24 inches long. To help estimate the amount of materials needed for the job, draw a plan view of the ceiling.

2 Locate the Benchmarks. The key to success is making sure you install the ceiling level across the entire room. Existing floor and ceiling surfaces may not be level, so never use them as reference points for measuring. Instead, establish benchmarks on the walls at every corner using a water level to ensure that each benchmark is precisely located. Benchmarks can be placed at any height, but a 60-inch height is most convenient.

Future measurements will be taken from these benchmarks.

3 Determine the Ceiling Height. The standard ceiling height is 96 inches. Ninety inches is the minimum height for lighting in a suspended ceiling. Once you've determined the ceiling height, measure up from the benchmarks to locate the position of the wall molding. Snap a chalk line on the walls around the perimeter of the room. Establish this line where you'll locate the top edge of the molding; the chalk marks won't be visible after the ceiling molding is in place.

4 Install the Wall Molding. Nail molding to the walls with 6d box nails, making sure each nail penetrates a stud. Use aviation snips to miter the molding at inside and outside corners. When cutting the wall molding to length, remember to account for the thickness of the adjacent wall angle. Butt lengths of wall molding where they meet mid-wall.

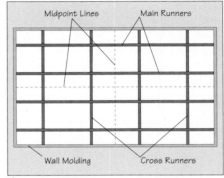

1 Make a plan-view drawing to determine the number of cross runners and ceiling panels you'll need, as well as the lineal feet of the main runners.

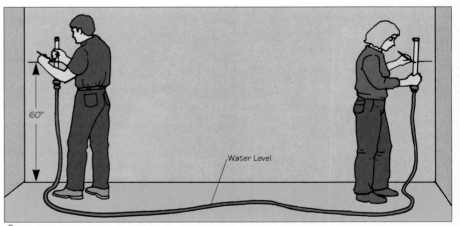

2 Use a water level to establish reference marks at each corner of the room.

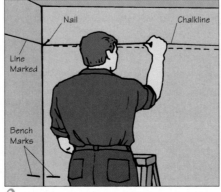

3 Measure up from the marks to locate the ceiling height, then snap a chalkline between the new marks.

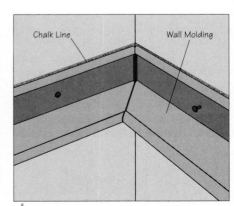

4 Use the chalkline as a placement guide to nail the wall molding in place.

5 Establish the Centerlines. Measure the length and width of the room and divide these measurements in half to get the center point of each wall. Use layout strings, stretching them tightly, to connect opposing midpoints. Check the intersection of the two strings to make sure they're square to each other. If not, adjust one or the other until they're square. It's easier to adjust layout strings when they're attached to nails that can be wedged behind wall molding.

6 Adjust the Layout. Plan the layout of ceiling panels to minimize the need for small pieces around the border of the ceiling. Doing so creates a better looking job. If the border tiles will be less than half the width of a field tile, adjust the layout one way or the other.

7 Install the Guidelines. Plan to install the first main runner approximately parallel to the wall and at a distance from the wall that's equal to the width of the border units. Make measurements from the centerlines rather than from the wall itself because the wall might not be square. Stretch a layout string between the wall moldings at these points.

8 Attach the Hanger Wires. Start with the joists at either end of the ceiling. Install a screw eye or a fastener supplied by the ceiling manufacturer into every fourth joist directly above the layout string. Twist a piece of hanger wire through each screw eye so that it hangs about 6 inches below the layout line. Cut a main runner to length and hang it from the wires so that it's just barely above the layout string. Twist the wires to secure the runner in position.

9 Install the Cross Runners. Slip the first cross runner between the main runner and the wall molding, locking it into the main runner's prepunched holes. Install the next main runner by using cross runners to gauge its spacing. Continue to work across the room until all of the runners have been installed.

5 Use a square to check the angle that's formed where the layout strings meet. Adjust the strings so they're at 90°.

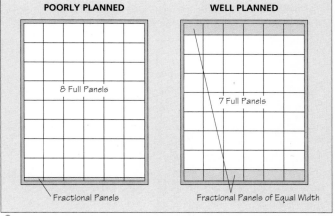

6 Adjust the layout so that border panels on opposite sides of the ceiling are the same size.

7 Stretch a layout string between opposite wall moldings. The layout line acts as a guide to the height of the first main runner.

8 Attach screw eyes directly above the layout string and loop a hanger wire through each screw eye. Secure the main runner to the wire.

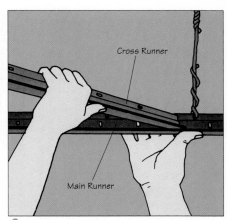

9 Install cross runners into slots in the main runner. Maintain the proper spacing according to the size of the ceiling panels used.

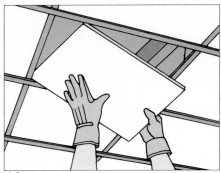

10 *Angle the ceiling panels through the grid and set them into place. Wear lightweight gloves to avoid marring the finished surface of the panels.*

10 Place the Ceiling Panels. Set each ceiling panel into place by turning it at an angle and pushing it into the grid of runners. Use a utility knife and a straightedge to cut the panels at the borders as needed. When handling the panels, wear clean, lightweight gloves so you don't smudge the finished surfaces.

Working around Obstructions

1 Hang Runners below the Obstruction. If a pipe or duct intrudes below the level of the ceiling, you can box it in with pieces of the grid system. You'll need U-shaped channel molding and extra wall molding for this job. Include the box in original layout plans and leave the ceiling open for now. Use the aviation snips to cut 90-degree notches in lengths of main runner, then bend the runners at these points to form the "ribs" of the box.

2 Attach the U-Channel. Fasten U-channel to the ceiling grid along the length of the obstruction. Then drill pilot holes and use a pop riveter to attach the box ribs to the U-channel with rivets. Pop riveters are available at hardware stores.

3 Install the Panels. Connect the ribs with lengths of ceiling molding and cut ceiling panels to fit the box as needed. Install the vertical panels first; they're locked in place when you install the horizontal panels. Use hanger wire as needed to provide additional support for the box.

4 Work around Posts. There are two ways to deal with posts that penetrate suspended ceilings: Either cut a panel in half and shape the pieces to fit around the post, or use additional runners to box-in the post. If you cut a panel in half, the seam will probably be unobtrusive and won't need extra support.

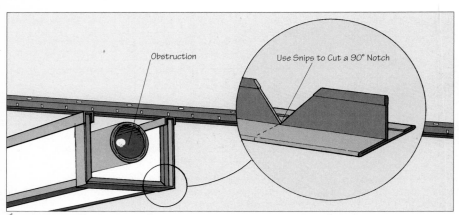

1 *Cut 90° notches in the spine of main runner stock, then bend the runner at the notches to form a U-shaped rib that surrounds the obstruction.*

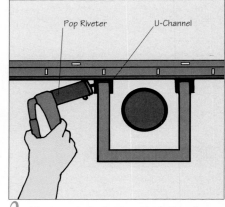

2 *Use a pop riveter to attach the ribs to the U-channel.*

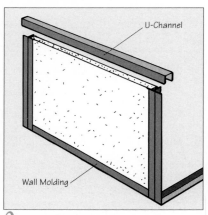

3 *Use wall molding to connect the ribs, then install the ceiling panels.*

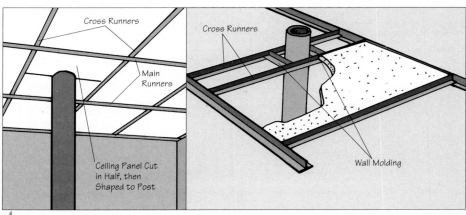

4 *To work around a post, slice a ceiling panel in half and cut each half to fit the post, or add runners and small pieces of ceiling panel to enclose it.*

Flooring, Trim, and Molding

10

Finish flooring and trim work are among the last tasks undertaken in any remodeling project. Because everything at a construction site seems to wind up on the floor eventually, it's best to wait until all the messy jobs are done before putting down the flooring. Trim and molding, especially unpainted trim work, is best applied after the walls and ceilings have been decorated. If you're planning on painting the molding, apply it just prior to painting.

Finish Flooring

You have a lot of choices when it comes to the kind of floor with which you'll finish your new attic, basement, or garage conversion. Since the space is likely to be a general living or sleeping area, you'll most likely choose from among wood, carpet, and vinyl flooring. Hardwood is the most elegant and expensive. It's also durable. Carpeting may be the most comfortable, though it can cause problems for those with dust-mite allergies and chemical sensitivities. Sheet vinyl is versatile and easy to lay; however, you'll most likely have to install an underlayment first.

Installing Hardwood Flooring

Putting in a hardwood floor is a job that requires a thorough knowledge of the process and a certain expertise in handling wood. If you're doing this project for the first time, study the procedures detailed in a hardwood flooring book. You may even wish to discuss the project with a lumber dealer. Should you already have hardwood floors in the rest of your house, be sure the new flooring for your conversion matches.

Assuming your house's structural members are sound and sturdy, you must adequately prepare the subfloor to receive the new flooring. A wooden floor, whether made from one-by lumber or plywood, makes a good subfloor if there are no seriously damaged sections. Drive down all nails until they're flush, correct any bowed boards, and replace badly warped or split boards or plywood panels.

Concrete makes an acceptable subfloor if it's dry. A moisture barrier—a thin sheet of polyethylene under 1x4 or 2x4 sleepers—will keep out dampness that could rot the floor (see "Installing an Insulated Subfloor," page 73). When ordering boards, judge the quality by standards set by the National Oak Flooring Manufacturers Association. In order of decreasing quality they are: clear, select, No. 1 common, and No. 2 common. The standards are determined by color, grain, and imperfections such as streaks and knots.

When ordering ¾x2¼-inch boards, multiply the number of square feet in the room by 1.383 to determine the amount of board feet you'll need, including wastage. For other size boards, ask your dealer how to compute the quantity. When you order the flooring, ask about renting a power nailer to aid in installation.

Difficulty Level:

Tools and Materials

- ☐ Basic carpentry tools
- ☐ Felt building paper, 15-pound
- ☐ Staple gun
- ☐ Chalkline
- ☐ Electric drill with assorted bits
- ☐ Flooring nails
- ☐ Power nailer (rented)
- ☐ Reducer strip
- ☐ Enough flooring to cover the area you're working on

1 Lay the Building Paper. Tack down any loose boards in the subfloor and set all exposed nailheads. Lay a covering of 15-pound asphalt-saturated felt building paper over the subfloor. Overlap the seams slightly and cut the edges flush with the walls. Nail or staple around the edges of each sheet. When the paper is in place, use chalk to mark the positions of the joists.

2 Lay Out the Work Lines. After you've laid the asphalt-felt building paper and marked the joist lines, it's important to lay out work lines based on either a wall that's square or the center of the room. Find the midpoints of the two walls that are

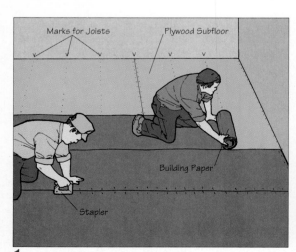

1 Cut building paper to fit closely around obstructions, tack it down, and mark the joist locations.

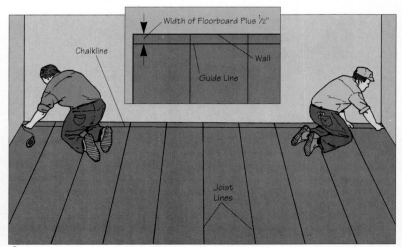

2 Measure from the centerline to establish a work, or guide, line parallel to the centerline.

parallel to the joists and snap a chalk line between them. From this line, measure equal distances to within about ½ inch of the end wall where you'll begin laying boards. Snap a chalk line between these two points and let this be your guide line for the first course of boards, regardless of how uneven the wall behind it may be. Any gap between the first course and the wall can be filled with boards trimmed to fit or covered with the baseboard and shoe molding.

3 Align a Starter Course. Along the work line that's drawn ½ inch from the wall, lay out the starter course (the first row of boards) the full length of the wall. Drill holes along the back edges of the boards and over the joists, slightly smaller than the nails, and face-nail the flooring.

4 Nail through the Tongue. Drill holes through the tongue of the first course of boards into the joists, then drive in finishing nails and set them. The first two rows of boards will be too close to the wall to use a power nailer.

5 Lay a Field. Lay out several courses of boards the way they'll be installed. Plan as much as six or seven rows ahead. Stagger the end joints so that each joint is more than 6 inches from the joints in the adjoining rows. You may have to cut pieces to fit at the ends of each row. Try to fit the pattern so that no end piece is shorter than 8 inches. Leave ½ inch between the ends of each row and the walls. When you've laid out a field of rows, begin to fit and nail.

6 Fit and Nail the Rows. As you lay each row, use a scrap of board as a tapping block. To keep from marring the board with the hammer when you nail, don't hammer nails flush into the tongue. Instead, leave the nailhead exposed, place the nail set sideways over it, and drive the nail home by hammering the nail set. Then use the tip of the nail set to countersink the nail into the board. When you reach the third row or so of boards, you'll have enough room to use a rented power nailer. Begin about 2 inches from the side wall and slip the power nailer onto the tongue of the last board laid.

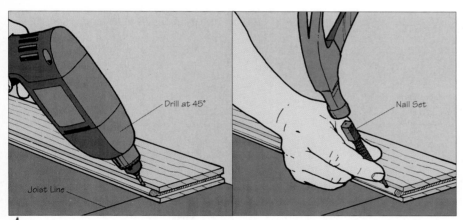

3 Position the first course of boards on the guide line and predrill them. Face-nail the boards in position and set the nails.

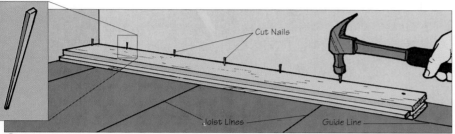

4 The tongue is fairly delicate so predrilling is advisable. Drill nailholes at the places marked for joists.

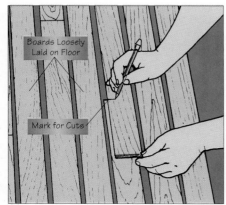

5 Plan the pattern of the boards by laying out several courses.

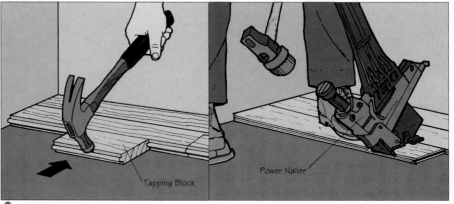

6 Use a tapping block to fit the boards tightly together and protect the tongue of the board. After two rows or so, you can use a power nailer.

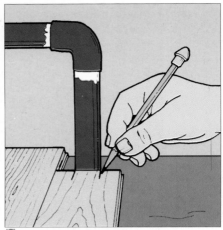

7 *Lay flooring up to obstacles so that you can slip the board to be cut alongside for measuring.*

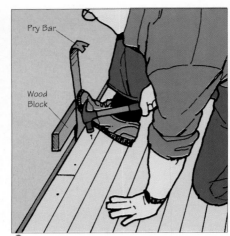

8 *Cut the reducer strip to the width of the doorway opening, and butt the flooring against the strip.*

9 *Use a pry bar between the wall and the last course to wedge the floorboards into position.*

Reversing Direction

If you intend to lay boards in hallways or closets that open off the room, you'll have to butt two boards, groove end to groove end, at the transition point. To reverse tongue direction, place a slip tongue (available from flooring dealers) into the grooves of the last course of boards nailed down, then set the grooves of the boards that will reverse the tongue direction over the slip tongue. Nail the reversed boards in place, driving the nails through the tongues, and proceed as usual.

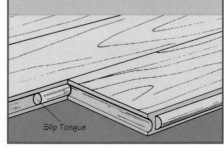

Hold the new board in position by placing your heel over it. Strike the plunger with a rubber-headed mallet hard enough to drive a nail through the tongue and into the floor. Drive a nail into each joist and into the subfloor halfway between joists.

7 **Cut around Obstacles.** When you come to an obstacle such as a radiator or a corner, test-fit the boards, measuring carefully. Make a cardboard template, if necessary, to transfer the cut onto the board. Decide whether you should save the tongue or groove. Clamp the board to a workbench and cut to fit with a handsaw or circular saw.

8 **Finish the Doorways.** To finish a doorway where the new floor will meet a floor that's lower, install a clamshell reducer strip by facenailing it. (The name comes from its rounded top, which resembles half of a clamshell.) The reducer strip is made so that one side will fit over the tongue of the adjoining board. The strip can also be butted to meet boards that run perpendicular to the doorway.

9 **Set the Final Board.** For gaps of more than ½ inch between the final board and the wall, remove the tongue sides of as many boards as you need, cut them to width, and wedge them into place with a pry bar. Hold them tightly with the pry bar by placing your foot on the bar while its hooked end pulls the filler board up tightly against the last board. Facenail these last boards, then install the baseboards and shoe moldings.

Installing Laminated Wood Floors

In addition to solid-wood strips and planks, wood flooring comes in an easy-to-install prefinished laminated version. As its name suggests, prefinished laminated wood flooring comes with a factory-applied, finished topcoat and is manufactured by cross-laminating layers of wood veneer. The ½-inch-thick flooring comes in strips 2 to 4 inches wide or planks more than 4 inches wide. The planks are often made of a number of narrow strips assembled together to resemble strip flooring when installed. In both forms, strip and plank, the flooring locks together with tongue-and-groove edges.

Because it's made from layers of wood plies, laminated flooring is more stable than solid wood and can be installed where ordinary wood flooring might have problems—say below grade and directly over concrete. This stability also enables the flooring to be installed as a floating floor system over a ⅛-inch thick layer of high-density foam underlayment. The tongue-and-groove joints are glued; no nails are used in the installation. Laminated flooring may be laid over any smooth, dry, level subfloor.

Difficulty Level: 🔨 *to* 🔨🔨

Tools and Materials

- ☐ Plastic (6 mil thick) as a vapor barrier when installing over concrete
- ☐ Foam underlayment, ⅛ inch thick
- ☐ Duct tape
- ☐ Laminated flooring
- ☐ White glue
- ☐ Hammer
- ☐ Measuring tape
- ☐ Pencil
- ☐ Handsaw or circular saw

1 Prepare the Floor. Roll out the foam underlayment and cut it to fit the room. Butt the joints and seal them with duct tape. If you're installing the floor over concrete, first lay a vapor barrier of polyethylene sheeting at least 6 mil thick. Overlap the vapor barrier joints by at least 6 inches and tape them with duct tape.

2 Lay the Planks. Leave a ½-inch expansion gap around the perimeter of the room. Start the installation on the longest wall of the kitchen, with the panel's groove facing the wall.

3 Join the Planks. Try to work in room-length runs. Join the planks by running a bead of white glue in the bottom edge of the groove as you install each succeeding section.

4 Finish the Job. Tap the sections together tightly with a hammer and a scrap piece of flooring as a hammering block. Mark, cut, and install planks in irregular areas just as you would a normal wood floor.

Installing Carpet

Standard wall-to-wall carpeting is installed on a pad and secured by tackless strips around the perimeter of a room to create tension. Before ordering carpet, prepare a scale drawing on graph paper, letting each square equal 1 foot. Mark all doors, alcoves, obstacles, and other features in the room you intend to

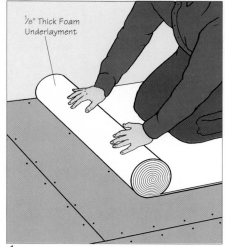

1 *Roll out foam underlayment and trim it to fit the room. Install a vapor barrier if you're putting the floor over concrete.*

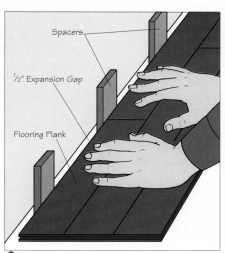

2 *Leave an expansion gap of ½ in.; begin laying the planking with the groove side against the wall.*

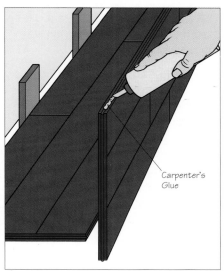

3 *Squeeze carpenter's glue into the bottom of each groove, then join the panels, tongue to groove.*

4 *Tap the joint tight along the entire run of planking with a hammer and a block made from scrap flooring.*

carpet. The more accurately you make the scale drawing, the easier it is for your carpet dealer to recommend the type and amount you need.

If you can't carpet your entire room with one piece, you'll have to allow for a seam. Seams are weak spots in the carpet and should not be placed where traffic flow is heavy. Sometimes seams are visible, depending on the kind of carpet you choose, and if possible should be laid away from the primary visual focus in the room. Seams are less visible when

they run parallel to prevailing light, so orient them toward the room's primary source of light—usually the wall with windows.

Prepare the subfloor before laying carpeting by nailing down loose boards, fixing squeaks, and replacing any rotten or damaged boards. If you intend to lay carpet on concrete, the concrete must be dry and not subject to sweating. Lay down a vapor barrier as a preventive measure. If the slab gets damp, cover it with a sealer before putting down the vapor barrier.

To lay carpet successfully, you'll need to rent two tools for stretching the carpeting material: a knee kicker and a power stretcher, available at your carpet dealer. To seam the carpet, you'll have to rent or borrow a seaming iron. You may also want to rent a carpet cutter.

Difficulty Level: 🔩🔩 to 🔩🔩🔩

Tools and Materials

- ☐ Hammer
- ☐ Measuring tape
- ☐ Handsaw
- ☐ Hacksaw
- ☐ Wood or cardboard spacer
- ☐ Padding
- ☐ Staple gun
- ☐ Utility knife
- ☐ Carpeting
- ☐ Chalkline
- ☐ Seaming tape
- ☐ Seaming iron
- ☐ Knee kicker (rented)
- ☐ Power stretcher (rented)
- ☐ Wall trimmer (rented)
- ☐ Putty knife (or screwdriver or trowel)

1 Install Tackless Strips and Binder Bars. Install wood tackless strips around the perimeter of the room with the tacks facing the walls. Cut the strips with a handsaw or chisel and nail them down a distance from the wall equal to about

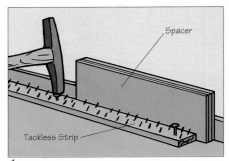

1 Attach tackless strips ⅔ the thickness of the carpeting away from the wall, with the nails facing the wall.

two thirds the thickness of the carpet. Use a cardboard or wooden spacer to place strips evenly. Attach binder bars, or standard metal edging, in doorways and other openings where the carpeting ends without a wall. The edging in the doorway should be placed directly under the door when the door is closed.

2 Fit the Pad. Cut the padding so that it covers the entire floor. Butt pieces evenly at any seams. If the padding has a waffle pattern, that side goes up. Staple the padding at 6-inch intervals around the perimeter of the room and any other places where it might slip. Ask a carpet dealer what he or she suggests for nonwooden floors.

3 Trim the Pad. After you've fastened the padding securely around the room, trim the overlap with a sharp utility knife just along the inside of the tackless strip. Leave

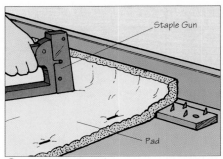

2 The pad should cover the entire floor, overlapping the tackless strips around the walls. Butt pieces to fill the area.

a ⅛- to ¼-inch gap between the tackless strip and the pad. If the padding is urethane or rubber, tilt the knife slightly away from the wall to create a beveled edge to prevent the pad from climbing.

4 Make the Rough Cuts. Roll out the carpet in a clean, dry, flat area. Allow at least 3 inches' overlap for the perimeter of the room and for any seams. Cut-pile carpeting should be cut from the back. First notch the ends where the cut will begin and end; then fold the carpet over and snap a chalk line on the back between the notches. Use white chalk if possible. Cut along the line with a utility or carpet knife, taking care to cut only the backing. If you have loop-pile carpeting, cut it from the front. Snap the chalk line and cut the carpeting with a utility knife. Your dealer will be able to tell you what kind of carpeting you have.

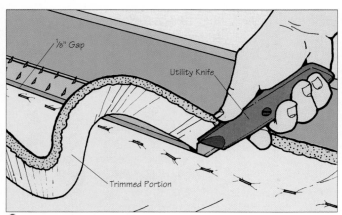

3 After stapling around the perimeter, trim the pad against the tackless strips with a utility knife.

4 How you cut carpet depends on the type: Cut-pile is cut from the back; loop-pile, from the front.

5 **Seam the Carpet.** To cut a seam, place one piece of carpeting over the other so the overlap is about 1 inch. Use the top piece as a guide to cut the bottom piece. Make sure the two pieces butt tightly, then insert a length of hot-melt seaming tape halfway under one piece of carpet. Make sure the adhesive side is up and the printed center is aligned with the edge of the carpet. Warm up a seaming iron to 250 degrees F. Hold back one edge of the carpet and slip the seaming iron under the edge of the other piece. Hold the iron on the tape for about 30 seconds, then slide it slowly along the tape while you press both halves of the carpet down behind it onto the

heated adhesive. Go slowly and be sure the two edges butt tightly. If the edges tend to pull away from each other, pull them together and place a heavy object on them until they have time to bond to the tape. Let the seam set.

6 **Stretch the Carpet.** Using a knee kicker, walk around the room and shift the carpet so it lies smoothly. Trim the edges to overlap the tackless strip by 1 or 2 inches. Make incisions for corners and cut around grates and other obstacles. Starting at a corner of the room, place the knee kicker about 1 inch from the tackless strip, and at a slight angle to one of the walls. Bump the kicker with your knee

so it moves the carpet and hooks the backing on the strip. Experiment with the power stretcher to learn how much "bite" is needed to grip the carpet and stretch it sufficiently. Pull the carpet taut with a minimum of force so it doesn't tear.

7 **Follow the Stretching Sequence.** The drawing on the next page shows the correct sequence for hooking and stretching the carpet. Follow the sequence of steps as you work to find the correct placements for the knee kicker and power stretcher. You'll use two techniques: hooking with the knee kicker in corners and along edges, and stretching with the power stretcher.

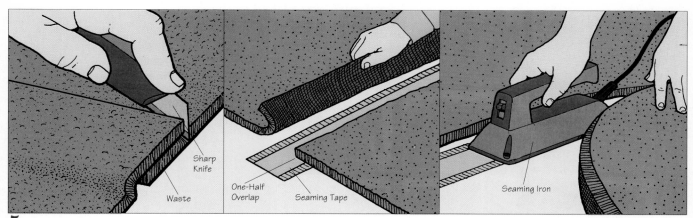

5 Cut tight seams by overlapping the fitted piece on the piece to be cut; use the top piece as the cutting guide (left). Slip a piece of seaming tape, cut to length, under the cut so that the tape is half under each side (middle). Lift the carpet back and run a 250°-F hot seaming iron over the tape. Press the carpet into the heated adhesive (right).

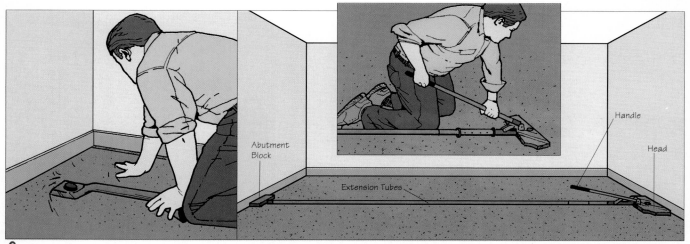

6 Set the end of the knee kicker 1 in. from the tackless strip and jam the carpet up over the strip (left). A power stretcher needs enough "bite" to move the carpet (right), but mustn't tear the carpet.

1. Hook corner A.
2. Stretch toward corner B and hook.
3. Hook edge A-B while keeping tension on corner B.
4. Stretch toward corner C and hook.
5. Hook edge A-C while keeping tension on corner C.
6. Stretch toward corner D and hook.
7. Hook edge C-D while keeping tension on corner D.
8. Stretch toward edge B-D and hook.

8 **Finish the Job.** The last step is to trim the carpeting between the wall and the tackless strip. Use a rented wall trimmer to make the job easy. If you can't get a wall trimmer, a utility knife will do. First adjust the trimmer to the thickness of the carpet. Slice downward into the carpeting at a 45-degree angle, leveling it out when you reach the floor. Leave just enough edging to tuck down into the gap between the strip and the wall. Use a putty knife, trowel, or screwdriver to push the edge of the carpet into the space between the tackless strip and the wall. If the carpet edge bulges, trim it a bit.

Lastly, with a block of wood and a hammer, clamp the carpet to the binder bars at the doorways and any other place you may have installed them.

Installing Resilient Sheet Flooring

Resilient sheet flooring can be installed over a wood subfloor or concrete floor. It's best if a wooden subfloor is covered with underlayment, as any gaps or imperfections in the subfloor boards will cause the material to bubble and not lay flat. Large cracks in, or uneven portions of, concrete slabs must be repaired to render a smooth surface.

Difficulty Level:

Tools and Materials

☐ Plywood, ¼- or ½-inch
☐ Fasteners (box nails, screws, or staples)
☐ Graph paper
☐ Pencil
☐ Template paper
☐ Resilient flooring
☐ Linoleum knife
☐ Heavy scissors
☐ Vinyl adhesive
☐ Seam sealer
☐ 20- to 24-inch 2x4
☐ Metal straightedge
☐ Linoleum roller (rented)
☐ Manufacturer-recommended solvent

1 **Lay the Underlayment.** An underlayment is necessary to eliminate irregularities in a subfloor over which resilient flooring will be

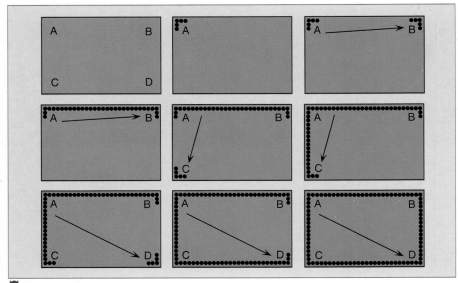

7 Carpeting should be stretched in the sequence shown above. The circles indicate the points of attachment, and the arrows indicate the direction of stretching.

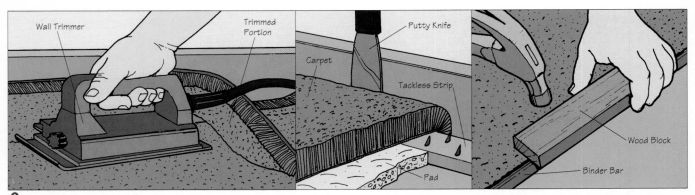

8 Cut the carpet at the wall with a rented wall trimmer or a utility knife if a trimmer isn't available. Leave enough carpet to be tucked against the wall (left). Push the carpet down with a screwdriver or putty knife between the tackless strip and the wall (middle). Slip the carpet into binder bars at doorways and close the bars with gentle taps on a wood block (right).

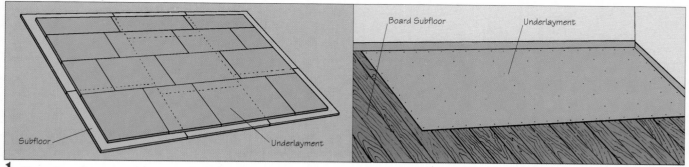

1 To make a floor sound, arrange underlayment over paneled subflooring so that its seams never fall directly over seams in the subflooring. For a board subfloor, lay the first panel of underlayment across the direction of the boards in the subfloor. If the end of the panel falls directly over a seam in the floor, fit it so that it falls in the middle of a board.

laid. Underlayment is either ¼-inch or ½-inch plywood. There are three methods of fastening underlayment to subfloors:

▶ Coated Box Nails. This kind of nail has a sheath of resin that melts with the heat of the friction of being driven into wood. The resin rehardens when the nails are in place, holding securely.

▶ Screws. Use care when driving screws to get their heads flush to or below the surface.

▶ Staples. Many professional home builders use staples driven by a power staple gun. You can rent the same kind of gun from a tool center.

With grade A-C plywood, lay the grade-A side up. Be sure that the panels of underlayment always span the seams of the subfloor. If the edge of the first panel of underlayment falls directly over a seam, cut it so that it and subsequent panels don't hit the seams. Locate the floor beams in the subfloor by the nailheads and extend their lines onto the underlayment. Drive 8d coated box nails (or staples or screws) every 4 to 6 inches along the floor beams, ⅜ inch from the edges of the underlayment.

2 Plan the Floor. Design the floor by using a piece of graph paper to map out the shape and dimensions as accurately as possible. Make a scale drawing to include irregularities like closets, alcoves, fireplaces,

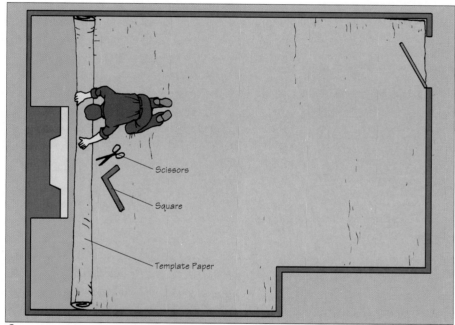

2 Unroll wide flooring paper or store-bought template paper to make a full-size template of the floor in the room. When you make the template, determine where the seam will go, based on the pattern of the flooring and the traffic in the room.

and doorways. If your floor is irregular, make a full-size template to use as a guide for cutting the sheets. Some manufacturers sell template kits, or you can make your own template with flooring paper.

3 Make the Rough Cuts. Take your floor plan to a local dealer and have him or her make the rough cuts. If you do plan to cut the flooring yourself, ask whether to reverse the sheets at the seam so the design falls into place. Use a linoleum knife and heavy scissors to cut the most intricate piece first, measuring so it's

3 inches oversize on all sides, including the seam. If you're using adhesive, spread it on the floor for this piece, stopping about 10 inches from the seam. Position the flooring. Then cut the second sheet so it overlaps at the seam at least 2 inches. Spread the adhesive over the rest of the floor, stopping 2 inches from the edge of the first sheet. Position and align the second piece carefully. Then cut half-moon shapes at the end of each seam so the ends butt the walls. With a straightedge and utility knife, cut through both sheets at the point where the seam will be. Lift up both

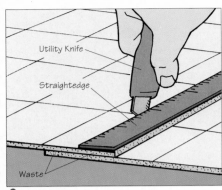

3 To make a perfect seam, install the two parts of the flooring, overlapping and cutting them as shown.

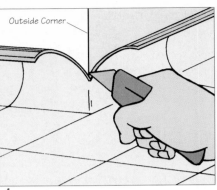

4 To trim an outside corner, start at the top of the flooring where it overlaps the corner and cut down to the floor.

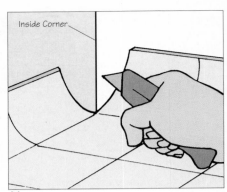

5 To trim an inside corner, cut the flooring in V-shaped sections down the corner until the flooring can lie flat.

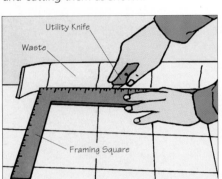

6 When trimming along a wall, use a straightedge to cut away any excess flooring.

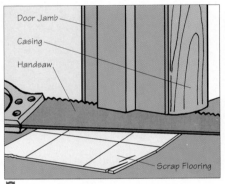

7 Cut under a door jamb as shown, resting the saw on a piece of new flooring used as a guide.

8 Use a linoleum roller or large rolling pin to flatten the floor and force out any air bubbles.

halves and apply adhesive. Clean the seam and use the recommended seam sealer for your flooring.

4 Trim the Outside Corners. Trim an outside corner by cutting straight down the curled-up flooring. Begin at the top edge and cut to where the wall and floor meet.

5 Trim the Inside Corners. Trim an inside corner by cutting the excess flooring away with increasingly lower diagonal cuts on each side of the corner. Gradually, these cuts should produce a wide enough split for the corner to wedge through and the flooring to lie flat around it.

6 Trim along the Walls. Remove the curled up flooring at the walls by pressing it down with a long 20- to 24-inch piece of 2x4. Press the flooring into the right angle, where the wall and floor

meet, until it begins to develop a crease at the joint. Then position a heavy metal straightedge into this crease and cut along the wall with a utility knife, leaving a 1/8-inch gap between the edge of the flooring and the wall. This is necessary for the material to expand without buckling.

7 Cut under the Door Jambs. The best way to have the flooring meet a door jamb is to cut away a portion of the jamb at the bottom so that the flooring will slide under it. Trim the flooring to match the angles and corners of the door jamb, overcutting about 1/2 inch for the edge to slip under the jamb.

8 Finish the Job. When cleaning up, avoid damaging the flooring by using a solvent recommended by the manufacturer. It's important to clean up any adhesive that may have spilled or oozed up onto the surface.

Then roll the flooring so that it sets firmly and flatly in the adhesive. You can use a rented linoleum roller or lean heavily on a large rolling pin and work your way across the floor. Start at the center of the room and roll firmly to remove air bubbles.

Trim and Molding

With any attic, basement, or garage conversion, it's important that trim and molding match those accents throughout the rest of the house. Uniform molding is an important key to making any remodeling project look as though it were part of the original house design.

Trim Versus Molding

Trim and molding come in a variety of shapes, sizes, and species of wood. "Trim" is the name given to wood that's rectangular in cross section, with no embellishments. Trim is typically 1 inch thick or less. Part of the confusion is that the word "trim" is also used as a general description of any wood in a house that isn't structural lumber—baseboards, casing, even moldings. The professional who traditionally installs all this material, for example, is called a trim carpenter. The word "molding," on the other hand, refers specifically to thin strips of material, usually wood, that have been cut, shaped, or embossed in some way to create a decorative effect. Molding includes everything from simple quarter-rounds to elaborate crown molding.

Types of Molding

You'll find that the molding bins are stuffed with every conceivable shape or profile of molding, and nearly every one is available in several dimensions.

Base Molding. Baseboard protects the lower portion of the walls, and covers any gaps between the wall and the floor. Base shoe molding is used to conceal variations between the floor and the baseboard bottom.

Ceiling Molding. Cove covers the inside corners between sheets of paneling. It's also used for built-up crown molding. Crown molding is used for dramatic effect at the juncture of walls and ceilings.

Wall Molding. Wainscot cap can be used to cover the exposed end grain on solid-wood wainscoting or to finish off the top of flat base-

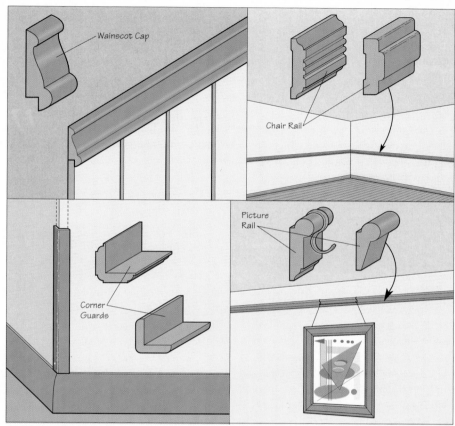

Wall Molding. *Wainscot cap and chair rail bridge differences between materials on a wall (top). Corner guards protect corners from damage (bottom left). Picture rail allows pictures to be moved without marring wall surfaces (bottom right).*

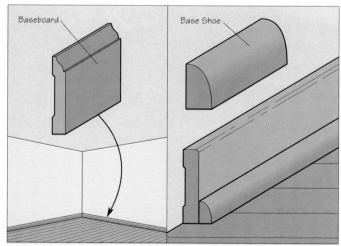

Base Molding. *Baseboard protects the bottoms of walls, and base shoe covers the edges of newly-installed flooring.*

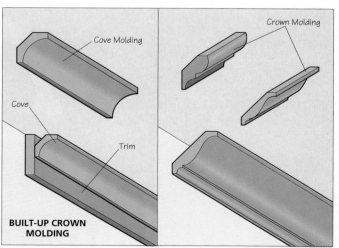

Ceiling Molding. *At the intersection of the wall and ceiling, cove or crown molding is installed.*

boards. Chair rail molding is installed at the height that protects walls from being damaged by chair backs. Chair rails are also used to cover the edges of wallpaper wainscoting. Corner guards protect the outside corners of drywall or plaster in high-traffic areas. Picture rail lets you add, move, and remove pictures without damaging the walls.

Casing. Casing conceals the gap between door and window jambs and the surrounding wall (see "Door Casing," page 120, and "Installing Casing around a Window," page 131). Common types of casing include clamshell, traditional, colonial, and ranch. Mullion casing is used as the center trim between

two or more closely spaced windows. Be sure the new casing is compatible with the existing window casing in style and thickness.

Coping a Baseboard

Cutting a coped joint is not difficult, but it does call for careful work and some patience. The value of the coped joint, as opposed to a miter, is that it won't as easily show a gap if the molding shrinks slightly.

Difficulty Level:

Tools and Materials

- ☐ Baseboard
- ☐ Backsaw
- ☐ Combination square
- ☐ Pencil
- ☐ Miter box or power miter saw
- ☐ Coping saw

- ☐ Round file
- ☐ Utility knife

1 **Install the First Piece.** Crosscut the first piece of molding and butt it into the corner.

2 **Make a Miter Cut.** Cut a 45-degree miter on the intersecting piece of baseboard, then use a square to draw a 90-degree pencil line at the top.

3 **Cut with a Coping Saw.** Use a coping saw to cut along the top front edge of the miter, following the profile line of the baseboard.

4 **Make the 90-Degree Cut.** Remove excess molding at the bottom to form a 90-degree angle. Test the fit by slipping the coped piece into place against the first piece of molding.

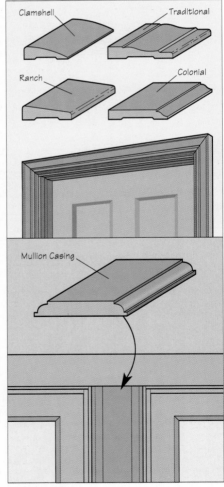

Casing. Window and door casing are among the most important trim elements for setting the style of a room.

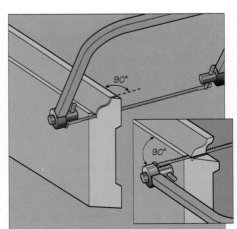

1 Butt the first piece of molding tightly into the corner.

3 Angling the saw slightly away from the edge of the miter, cut down to the front edge.

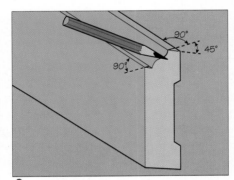

2 Pencil a 90° line on the top of the edge of the miter-cut baseboard.

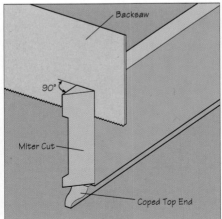

4 Remove the remaining 45° angle of the miter cut at the bottom of the molding with a miter saw.

5 **Make Trim Cuts.** Use a round file or utility knife to fine-tune the back side of the cut. Retest the fit until you're happy with it.

6 **Install the Second Piece.** The two cuts should produce a face that fits the contours of the piece to which it's butted.

Even a baseboard with a simple profile should be coped for the best fit. Note that where the two pieces meet at the top edge, a fine piece of wood will overlap from the coped piece to the butted piece. Take care during installation not to damage this fragile point.

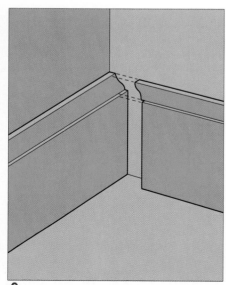

5 You may have to remove wood so the front edge of the cope fits properly.

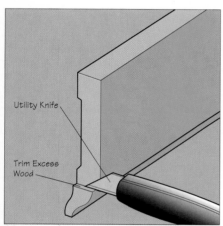

6 Well-cut contours fit together perfectly.

Installing Baseboard Molding

In many cases, especially when you're leaving it natural, installing base trim will be your last project. Base trim is laid so that each has a butt joint on one end and a cope or miter joint on the other. Once you get the feel for installing baseboard, the work goes quickly. All molding should be installed with care, but there's more allowance in baseboards because they're not as visible as other moldings. Measure all the pieces before cutting them, or hold each one against the wall and mark the cut.

Difficulty Level:

Tools and Materials

☐ Coped baseboard
☐ Backsaw
☐ Miter box or power miter saw
☐ Finishing nails, 8d
☐ Hammer
☐ Nail set
☐ Putty knife
☐ Wood putty

1 **Cut the Baseboard.** Cut the baseboard to rough lengths and distribute them around the room.

2 **Miter the Pieces.** Start with an outside corner, it there is one. Miter and fit the first piece, then tack it temporarily in place. Cut lengths of baseboard slightly long so that you have to bow them a little to get them into place. This force-fit will guarantee tight joints.

3 **Tack the Baseboard in Place.** Work your way around the room, tacking each length of baseboard into place temporarily.

4 **Nail the Baseboard.** After all the baseboards are in place and fitted properly, drive all the nails home and set them with a nail set. Fill the nailholes with wood putty.

1 At this point, the baseboard needs only to be cut to the approximate size.

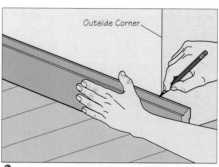

2 Mark the baseboard for the finish cut, making it slightly longer than needed.

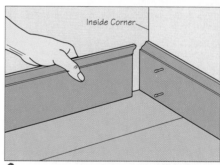

3 Use finishing nails to tack the baseboard in place.

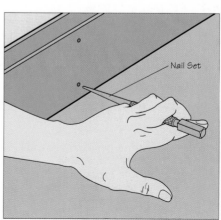

4 Holding the nail set like this makes it easier to drive nails close to the floor.

 Glossary

Bearing wall A wall that provides structural support to framing above, such as ceiling joists or roof members. Joists run at right angles to and rest on the top plate of a bearing wall.

Building codes Municipal rules regulating safe building practices and procedures. The codes generally encompass structural, electrical, plumbing, and mechanical remodeling and new construction. Inspection may be required to confirm adherence to local codes.

Butt joint A joint in which a square-cut piece of wood is attached to the end or face of a second piece.

Casing The trim that is used to line the inside and outside of a doorway or window frame.

Caulking A waterproof, adhesive filler material that remains flexible so that it will not pop or flake out of seams and cracks.

Chair rail A horizontal strip of molding mounted at the proper height and protruding enough to prevent the top of a chair back from touching a wall surface. See Wainscoting.

Chalkline A cord that is rubbed with or drawn through chalk and stretched taut between two points, just above a surface. It is pulled up in the center and released so that it snaps down, leaving a straight line marked on the surface between the end points.

Circuit The electrical path that connects one or more outlets and/or lighting fixtures to a single circuit breaker or fuse on the control panel.

Circuit breaker A protective device that opens a circuit, cutting off the power automatically when an overcurrent or short-circuit occurs.

Conduit Metal or plastic tubing designed to enclose electrical wires.

Cove molding A molding with a concave face used as trim or finish for interior corners.

Dado A square, U-shaped groove cut into the face of a board to receive and support the end of another board, such as the end of a shelf.

Dormer A window set upright in a sloping roof, and the roofed projection in which the window is set. This style provides increased headroom, natural light, and increased ventilation.

Drywall Also known as wallboard, gypsum board, plasterboard, and by the trade name Sheetrock, a wall-surfacing material composed of sheets of gypsum plaster sandwiched between a low-grade backing paper and a smooth-finish front surface paper that can be painted.

DWV (Drain, waste, vent system) The system of pipes and fittings inside the walls used to carry away plumbing drainage and waste.

Fish tape Flexible metal strip used to draw wires and cable through walls, raceways, and conduit.

Flashing Material used to prevent seepage of water around any intersection or projection in a roof, including vent pipes, chimneys, skylights, dormers, and roof valleys.

Furring Strips of wood attached to a wall to provide support and attachment points for a covering such as hardboard paneling.

Ground-fault circuit interrupter (GFCI) A safety circuit breaker that compares the amount of current entering a receptacle on the hot wire with the amount leaving on the white wire. If there is a discrepancy of 0.005 volt, the GFCI breaks the circuit in a fraction of a second.

Header A structural member that forms the top of a window, door, skylight, or other opening to provide framing support and transfer weight loads. Header thickness must equal wall width.

Jamb The inside face of a window or door.

Joint The space between the adjacent surfaces of two components joined and held together by nails, glue, cement, mortar, or other means.

Joint compound A premixed gypsum-based material with the consistency of mortar used to fill the seams in gypsum-board construction. Also called gypsum compound.

Joist One in a series of parallel framing members that supports a floor or ceiling load. Joists are supported by beams or bearing walls.

Kneewall A wall that extends from the floor of an attic to the underside of the rafters. Knee-walls are short (usually 48 inches high) and often non-bearing.

Load-bearing wall A wall that is used to support the house structure and transfer weight to the foundation.

Mastic The thick adhesive used to hold floor and wall tiles in place.

Miter A joint in which the ends of two pieces of wood are cut at equal angles (typically 45 degrees) to form a corner.

Molding Thin strip of wood that has a profile created by cutting and shaping.

Non-bearing wall An interior wall that does not provide structural support to any portion of the house above it. It usually runs parallel to ceiling joists, and can be removed without concern for supporting the overhead structure.

Paneling Planks or sheets used as a finished wall or ceiling surface; often with a wood or simulated wood finish.

Particleboard A structural sheet material composed of compressed wood chips, flakes, or small wood particles such as sawdust, held together with special glues.

Partition wall A wall that divides space but plays no part in a building's structural integrity.

Plywood Pieces of wood made of three or more layers of veneer joined with glue, and usually laid with the grain of adjoining plies at right angles. Almost always an odd number of plies are used to provide balanced construction.

Rabbet An L-shaped groove cut into the edge of a board to receive the edge of another board and form a corner joint.

Radon A colorless, odorless radioactive gas that comes from the natural breakdown of uranium in soil, rock, and water. When inhaled, molecules of radon lodge in the lungs and lead to an increased risk of lung cancer.

Rafters Dimensional lumber that supports the sloping roof of a structure.

Ridgeboard The horizontal framing piece to which the rafters attach at the roof ridge.

Ridge The horizontal line at which two roof planes meet when both roof planes slope down from that line.

Riser The vertical stair member between treads.

Rout The removal of material, by cutting, milling, or gouging, to form a groove.

Seepage The infiltration of water from outside to inside through cracks, seams, or pores.

Sheathing The wooden covering on the exterior of walls and the roof. Typically made of ½-inch construction-grade plywood; older homes may have shiplap boards or planks.

Shim A thin insert used to adjust the spacing between, for example, a floor and a sleeper laid over it.

Sistering The process of reinforcing a framing member by joining another piece of lumber alongside it.

Sleeper A strip of wood, usually a 2x4, laid flat over a floor to provide a raised, level base for a support member of a new floor above.

Soffit A short wall or ceiling filler piece. For example, the filler between the top front edge of a wall cabinet and the ceiling above.

Stringer Diagonal boards that support stair treads, usually one on each side and one in the middle of a staircase.

Stud Vertical member of a frame wall, placed at both ends and most often every 16 inches on center. Provides structural framing and facilitates covering with drywall or plywood.

Subfloor The surface below a finished floor. Usually made of sheet material like plywood.

Sump pump A device that draws water beneath the slab and pumps it away from the house.

Toenail Joining two boards together by nailing at an angle through the end, or toe, on one board and into the face of another.

Tread The horizontal boards on stairs, supported by the stringers.

Trim Unmolded strips of wood used alone or in combination with molding.

Trusses A roof-framing system with rafters supported by crossed webs.

Underlayment Sheet material placed over a subfloor or old floor covering to provide a smooth, even surface for a new covering. Underlayment is usually sheets of hardboard, particleboard, or plywood.

Vapor barrier Material used to block out the flow of moisture.

Wainscoting Paneling that extends 36 to 42 inches or so upward from the floor level, over the finished wall surface. It is often finished with a chair rail at the top. See Chair rail.

Index

Access
in attic conversions, 15–17; in basement conversion, 17–18, 38; in garage conversions, 18–19

Aesthetics in loft design, 20

Anchors, masonry, 100–102

Attic conversions
access for, 15–17; bathrooms in, 46, 152; building partition walls for, 80–84; ceiling joists in, 54; checking for trusses, 34; concealing heating ducts, 164; doors in, 84; dormers in, 89–98; eliminating moisture problems in, 55; family room/playrooms in, 23; floors for, 67–72; home office in, 28; insulation and ventilation for, 86–89; investigating problems in, 34–36; joists in, 67–70; walls in, 84–86; lighting for, 32; logistics planning for, 46; skylights in, 135–138; stairs for, 64–67; subflooring in, 70–72; uses for, 16–17

Balustrade, 62

Baseboard, 186–187. See also Base molding

Basement conversions
bathrooms in, 152; bedrooms in, 24–25; eliminating moisture problems in, 56; family room/playrooms in, 23–24; floors for, 72–74; framing in, 99–112; gaining access for, 17–18; heating system in, 55, 164; inspecting for problems in, 37–40; lighting for, 31–32; logistics planning for, 46–47; noise control in, 21; stairs for, 62–63; storage/utility area in, 27; suspended ceilings for, 170–174; types of walls in, 36–37; uses for, 17–18; windows in, 132–133; window wells in, 134–135; wiring for, 148–151; workshops in, 27

Base molding, 185
coping of, 186; installing, 187

Bathrooms
in attic conversions, 46, 152; in basement conversions, 152; in garage conversions, 152–153; planning for, 26, 153–154

Beams, 163–164

Bearing walls, 48

Bedroom, planning of, 24–25

Bids, 42

Boxing around posts, 162–163

Building codes, 10–11, 21, 26, 42, 43–44, 132

Carbon monoxide detector, 55

Carpentry tools, basic, 7

Carpet, installing, 179–182

Casing
door, 120–121, 186; windows, 131–132, 186

Catwalks in attic conversions, 46

Ceiling molding, 185

Ceilings, suspended, 170–174

Chimney problems in attic conversions, 36

Closets, planning of, 25

Collar ties, removing, 50

Concealing
beams, 163–164; ducts, 164; soil pipes, 164

Concrete floor
painting, 73; repairing cracks in, 72–73

Condensation
testing for, 38; eliminating, 56

Contractors, hiring, 42

Coping of baseboard, 186–187

Copper tubing, 154–156

Cost evaluation, 10–11

Demolition safety, 47

Dining room, 22

Doors
in attic conversions, 84; casing for, 120–121; framing, 115, 116–117; in garage conversions, 110–112; hanging, 116–117; hardware for, 117, 121–123; installing, 117–120; styles of, 114–116

Dormers
building gable, 96–98; building shed, 92–96; planning, 91–92; types of, 89–90

Drainpipe systems, 157–158

Driveways in garage conversions, 19

Drywall
in concealing beams, 164; finishing, 168–170; installing, 166–168; painting, 170; removing, 47–48; types of, 165–166; wallpapering, 170

Ducts, concealing, 164

Dust control, 27

Efflorescence, removing, 57

Electrical code, 43–44, 143, 151

Electrical systems. See also Lighting; Wiring
for home office, 28; installing receptacles, 145–146; planning for, 41; removing electric garage door opener, 51; safety in, 140; tools and materials, 8, 140, 141; wiring for, 140–148

Energy codes, 43

Carpet, installing, 179–182

Family room, planning of, 23–24

Fiberglass insulation, 105–108

Fire safety, 156

Fixtures. See Electrical systems; Lighting; Wiring

Flanged windows, installing, 128–129

Floor plans, 11
plumbing in, 153, 154; traffic patterns in, 11–13, 29

Floors
for attic conversions, 35, 67–72; for basement conversions, 72–74; for garage conversions, 75–78; installing carpeting for, 179–182; installing hardwood, 176–178; installing laminated wood, 178–179; installing resilient sheet, 182–184

Foundation cracks, checking basement for, 37–38

Framing
around posts, 161–162; in basement conversions, 99–112; for doors, 115, 116–117; in garage conversions, 99–112; removing, 48–49

Furniture clearance, 29–30

Gable-end windows, installing, 124–125

Garage conversions
bathrooms in, 152–153; concealing heating system in, 55; eliminating moisture problems in, 56; enclosing garage door opening, 110–112; floors in, 75–78; framing in, 99–112; gaining access for, 18–19; installing windows in, 129–130; lighting for, 32; logistics planning for, 47; noise control in, 21; removing garage door in, 51–53; skylights in, 135–138; storage/utility area in, 27; surveying for problems in, 40; suspended ceilings for, 170–174; uses for, 18–19; wiring for, 151; workshops in, 27

Ground-fault circuit-interrupter (GFCI) receptacle, 145

Hardware, door, 117, 121–123

Hardwood flooring, installing, 176–178

Headroom
in attic conversions, 34–35; in basement conversions, 38

Health hazard, radon as, 39

Heating systems
concealing, 55, 164; planning for, 40–41

Home office, planning, 27–29

Insect problems
in attic conversions, 36; in basement conversions, 39–40

Insulation
in attic conversions, 86–89; for garage conversions, 110; for masonry walls, 105–109

Joists
in attic conversions, 53–54, 68–70; in basement conversions, 40; in garage conversions, 78

Kitchen, planning of, 22

Kneewalls, 50, 84–86

Laminated wood floors, installing, 178–179

Lighting. See also Electrical systems
in attic conversions, 32; in basement conversions, 31–32; in garage conversions, 32; general requirements for, 30; types of fixtures, 31; for workshop, 27

Living room, planning of, 24

Locksets, 122–123

Lofts, uses for, 19–20

Logistical planning, 46–47

Masonry walls. See also Walls
fastening objects to, 100–102; insulating, 105–109; painting, 165; sealing, 56–58

Master bathrooms, 26

Master bedrooms, 25

Mechanical code, 43

Moisture problems
in attic conversions, 35–36, 55; in basement conversions, 38, 56; in garage conversions, 56

Molding, 185–186

Nails
drywall, 166; masonry, 102; removing, 50

National Electrical Code, 43, 143, 151

Noise control in room planning, 21

Non-bearing walls, 49
removing, 49–50

Overbuilding, 10

Painting
concrete floors, 73; drywall, 170; masonry walls, 58, 165

Parking, as concern, in garage conversions, 19

Partition walls. See also Walls
in attic conversions, 80–84; building, 102–105

Planning
importance of, 6–7; logistical, 46–47; room, 20–30

Playroom, planning of, 23–24

Plumbing, in wall removal, 49

Plumbing code, 43

Plumbing system, 151–157
basic tools for, 8; for bathrooms, 26, 152–153; concealing soil pipes, 164; copper tubing in, 154–156; drainpipe systems in, 157–158; planning for, 41–42; relocating existing lines, 156–157; sump pumps in, 59–60

Posts
boxing around, 162–163; framing around, 161–162; installing suspended ceiling around, 174; working with, 160–163

Preparation work, 45–60

Privacy
in bedroom design, 25; in master bathroom design, 26

Radon, 39

Rafters
in attic conversions, 35, 54, 93, 94–95; repairing, 53–54

Receptacles, installing, 145–146

Resilient sheet flooring, installing, 182–184

Roof, ventilating, 86

Room dimensions, 21

Room planning, 20–30

Safety
demolition, 47; electrical, 140; fire, 156; in lofts, 20

Sealing of masonry walls, 56–58

Seepage, eliminating, 56

Service corridors in concealing heating systems, 55

Siding, installing, 112

Skylights
installing, 135–136, 137–138; types of, 136–137

Snips, 171

Soil pipes, concealing, 164

Soldering of copper tubing, 154–156

Stairs
for attic conversions, 15–16, 63–67; for basement conversions, 62–63

Storage/utility area, planning of, 27

Subfloor
in attic conversions, 46, 70–72; in basement conversions, 73–74; in garage conversions, 77–78

Sump pumps, 59
installing, 59–60; types of, 59

Suspended ceilings, 170–174
anatomy of, 170–171; installing, 172–174; working around obstructions, 174

Time factor, 8

Tools and materials, 7–8

Traffic patterns
adding, to floor plans, 11–13; in furniture layout, 29

Trim, versus molding, 185

Trusses
checking for, in attic conversions, 34; in installing skylights, 135

Unused areas, making use of, 14–20

Utilities. See also Electrical systems; Heating systems; Plumbing systems
planning for, 40–42

Ventilation
in attic conversions, 35, 86–89; in basement conversions, 38

Wall molding, 185–186

Wallpapering drywall, 170

Walls. See also Masonry walls; Partition walls
closing foundation, 75–76; drywall for, 165–170; finishing, 165–170; painting, 165; removal of, 47–50; running cable through finished, 144–145; types of basement, 36–37

Waterproofing, 58–59

Windows. See also Skylights
in basement conversions, 132; boxing, 133; in dormers, 95–96; in garage conversions, 111–112, 129–130; installing casing around, 131–132; installing flanged, 128–129; installing gable-end, 124–126; installing new, 126–128; replacing wood, 133; types of, 123–124

Windowsills, beveling, 133–134

Window stool, installing, 130–131

Window wells, installing, 134–135

Wiring, 140–145. See also Electrical systems
for basement conversions, 148–151; choosing right cable in, 142; for electrical receptacles, 146; estimating needs in, 142; for garage conversions, 151; joining in, 141–142; for lighting fixtures, 146–148; relocating existing, 150–151; running cables in, 141, 144–145, 147–148; in wall removal, 49

Wood floors. See Hardwood flooring

Wood window, replacing, 133

Workshop, planning of, 27

Zoning restrictions, 29, 44

Metric Conversion Charts

Lumber

Sizes: Metric cross sections are so close to their nearest U.S. sizes, as noted at right, that for most purposes they may be considered equivalents.

Lengths: Metric lengths are based on a 300mm module, which is slightly shorter in length than an U.S. foot. It will, therefore, be important to check your requirements accurately to the nearest inch and consult the table below to find the metric length required.

Areas: The metric area is a square meter. Use the following conversion factor when converting from U.S. data: 100 sq. feet = 9.29 sq. meters.

Metric Lengths

Lengths Meters	Equivalent Feet and Inches
1.8m	5' 10⅞"
2.1m	6' 10⅝"
2.4m	7' 10½"
2.7m	8' 10¼"
3.0m	9' 10⅛"
3.3m	10' 9⅞"
3.6m	11' 9¾"
3.9m	12' 9½"
4.2m	13' 9⅜"
4.5m	14' 9⅓"
4.8m	15' 9"
5.1m	16' 8¾"
5.4m	17' 8⅝"
5.7m	18' 8⅜"
6.0m	19' 8¼"
6.3m	20' 8"
6.6m	21' 7⅞"
6.9m	22' 7⅝"
7.2m	23' 7½"
7.5m	24' 7¼"
7.8m	25' 7⅛"

Dimensions are based on 1m = 3.28 feet, or 1 foot = 0.3048m

Metric Sizes (Shown Before Nearest U.S. Equivalent)

Millimeters	Inches	Millimeters	Inches
16 x 75	⅝ x 3	44 x 150	1¾ x 6
16 x 100	⅝ x 4	44 x 175	1¾ x 7
16 x 125	⅝ x 5	44 x 200	1¾ x 8
16 x 150	⅝ x 6	44 x 225	1¾ x 9
19 x 75	¾ x 3	44 x 250	1¾ x 10
19 x 100	¾ x 4	44 x 300	1¾ x 12
19 x 125	¾ x 5	50 x 75	2 x 3
19 x 150	¾ x 6	50 x 100	2 x 4
22 x 75	⅞ x 3	50 x 125	2 x 5
22 x 100	⅞ x 4	50 x 150	2 x 6
22 x 125	⅞ x 5	50 x 175	2 x 7
22 x 150	⅞ x 6	50 x 200	2 x 8
25 x 75	1 x 3	50 x 225	2 x 9
25 x 100	1 x 4	50 x 250	2 x 10
25 x 125	1 x 5	50 x 300	2 x 12
25 x 150	1 x 6	63 x 100	2½ x 4
25 x 175	1 x 7	63 x 125	2½ x 5
25 x 200	1 x 8	63 x 150	2½ x 6
25 x 225	1 x 9	63 x 175	2½ x 7
25 x 250	1 x 10	63 x 200	2½ x 8
25 x 300	1 x 12	63 x 225	2½ x 9
32 x 75	1¼ x 3	75 x 100	3 x 4
32 x 100	1¼ x 4	75 x 125	3 x 5
32 x 125	1¼ x 5	75 x 150	3 x 6
32 x 150	1¼ x 6	75 x 175	3 x 7
32 x 175	1¼ x 7	75 x 200	3 x 8
32 x 200	1¼ x 8	75 x 225	3 x 9
32 x 225	1¼ x 9	75 x 250	3 x 10
32 x 250	1¼ x 10	75 x 300	3 x 12
32 x 300	1¼ x 12	100 x 100	4 x 4
38 x 75	1½ x 3	100 x 150	4 x 6
38 x 100	1½ x 4	100 x 200	4 x 8
38 x 125	1½ x 5	100 x 250	4 x 10
38 x 150	1½ x 6	100 x 300	4 x 12
38 x 175	1½ x 7	150 x 150	6 x 6
38 x 200	1½ x 8	150 x 200	6 x 8
38 x 225	1½ x 9	150 x 300	6 x 12
44 x 75	1¾ x 3	200 x 200	8 x 8
44 x 100	1¾ x 4	250 x 250	10 x 10
44 x 125	1¾ x 5	300 x 300	12 x 12

Dimensions are based on 1 inch = 25mm

For *all* of your home improvement and repair projects, look for these and other fine

Creative Homeowner Press books

at your local home center or bookstore...

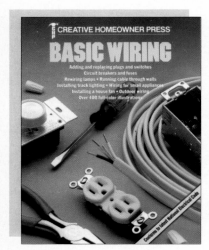

Basic Wiring (Third Edition, Conforms to latest National Electrical Code)

Included are 350 large, clear, full-color illustrations and no-nonsense step-by-step instructions. Shows how to replace receptacles and switches; repair a lamp; install ceiling and attic fans; and more.

BOOK #: 277048   160pp., 8½"x11"

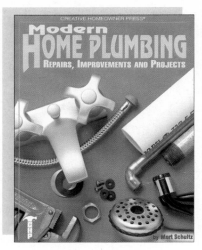

Modern Home Plumbing

Take the guesswork out of plumbing repair and installation for old and new systems. Projects include replacing faucets, unclogging drains, installing a tub, replacing a water heater, and much more. 500 illustrations and diagrams.

BOOK #: 277612   160pp., 8½"x11"

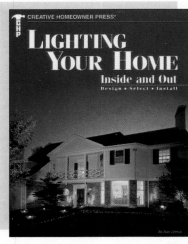

Lighting Your Home Inside and Out

This beautiful book thoroughly explains lighting design for every room in the house as well as outdoors. But this is much more than a design book. It is a step-by-step manual that shows how to install the fixtures that make their designs work.

BOOK #: 277583   160pp., 8½"x10⅞"

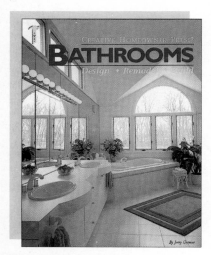

Bathrooms

Shows how to plan, construct, and finish a bathroom. Remodel floors; rebuild walls and ceilings; and install windows, skylights and plumbing fixtures. Specific tools and materials are given for each project.

BOOK #: 277053   192pp., 8½"x10⅞"

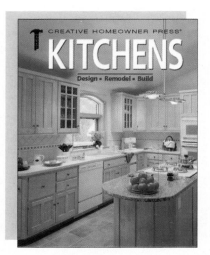

Kitchens: Design, Remodel, Build

This is the definitive reference book for modern kitchen design, with more than 100 full-color photos to help homeowners plan the layout and personalize the design. Step-by-step instructions and more than 470 colorful drawings illustrate basic plumbing and wiring techniques; how to finish walls and ceilings; how to lay flooring; and much more.

BOOK #:277059   192pp., 8½"x10⅞"

Working with Tile

Design and complete interior and exterior tile projects on walls, floors, countertops, shower enclosures, patios, pools, and more. Included are ceramic, resilient, stone, and wood parquet tile. 425 color illustrations and over 80 photographs show use, installation, repair, and maintenance.

BOOK #: 277540   176 pp., 8½"x10⅞"

For more information, and to order direct call 800-631-7795; in New Jersey 201-934-7100